T0218627

THE COVID-19 PANDEMIC

This book presents a comprehensive account of the COVID-19 pandemic, also known as the novel coronavirus pandemic, as it happened.

This volume examines the first responses to the COVID-19 pandemic, the contexts of earlier epidemics and the epidemiological basics of infectious diseases. Further, it discusses patterns in the spread of the disease; the management and containment of infections at the personal, national, and global level; effects on trade and commerce; the social and psychological impact on people; the disruption and postponement of international events; the role of various international organizations like the WHO in the search for solutions; and the race for a vaccine or a cure.

Based on new data and latest developments, the second edition of this volume explores the global spread of COVID-19 since 2019 and examines the emergence of the evolving coronavirus variants (Alpha, Beta, Gamma, Delta and Omicron). Further, it extensively discusses what we have since discovered on the disease, along with recent progress on treatments and vaccines.

Authored by a medical professional and an economist working on the frontlines, this book gives a nuanced, verified and fact-checked analysis of the COVID-19 pandemic and its global response. A one-stop resource on the COVID-19 outbreak, it is indispensable for every reader and a holistic work for scholars and researchers of medical sociology, public health, political economy, public policy and governance, sociology of health and medicine, and paramedical and medical practitioners. It will also be a great resource for policymakers, government departments and civil society organizations working in the area.

Tapas Kumar Koley is M.D. (internal medicine), senior specialist physician at Tilak Nagar Colony Hospital, South Delhi Municipal Corporation, New Delhi, India. He also works in the area of medicine and law related to medical negligence and has been involved in the training of doctors and lawyers on medical negligence. A few of his recent publications include *Medical Negligence and the Law in India* (2010) and *Handbook of Clinical Electrocardiography* (2009).

Monika Dhole is the joint director of Economic Affairs and Research Division with the Federation of Indian Chambers of Commerce and Industry (FICCI), New Delhi, India. She has worked with some of the most renowned research organizations in various capacities in India and has made research contributions to their publications and research projects over the years.

THE COVID-19 PANDEMIC

The Deadly Coronavirus Outbreak

Second Edition

Tapas Kumar Koley and Monika Dhole

Routledge
Taylor & Francis Group

LONDON AND NEW YORK

Cover image: wildpixel, Getty Images

Second edition published 2023
by Routledge
4 Park Square, Milton Park, Abingdon, Oxon, OX14 4RN

and by Routledge
605 Third Avenue, New York, NY 10158

Routledge is an imprint of the Taylor & Francis Group, an informa business

First edition published by Routledge 2021

British Library Cataloguing-in-Publication Data
A catalogue record for this book is available from the British Library

ISBN: 978-1-032-38289-0 (hbk)
ISBN: 978-1-032-38453-5 (pbk)
ISBN: 978-1-003-34509-1 (ebk)

DOI: 10.4324/9781003345091

Typeset in Bembo
by Apex CoVantage, LLC

The book is dedicated to our father, Late Mr B. N. Koley, and our mother, Mrs Kanan Koley.

CONTENTS

FIGURES

PREFACE TO THE FIRST EDITION

The ongoing COVID-19 pandemic is taking a heavy toll worldwide. It has caused death and debility throughout the world. It has disrupted the life and lifestyle of almost everyone. Almost no one has been left untouched. During this pandemic, another pandemic of information and misinformation is keeping pace with it, spreading fear and anxiety. Our profession has provided us with a rare opportunity to observe and feel the effects of the pandemic on everyday lives. This book is written to provide an exact picture of the destructive capacity of the novel coronavirus and at the same time remove the fear factor, by providing information to keep one safe from catching the infection.

The book is divided into nine chapters. Each chapter is written in plain language, avoiding technical jargon. Each chapter ends with references that will help an enlightened reader study further about the pandemic. Photographs and figures are provided to create a lasting impression. The statistical data provided have been collected from the World Health Organization (WHO), which is one of the most reliable sources of information about the COVID-19 pandemic. However, the pandemic is still evolving, and the data changes frequently, so there are several technical limitations to arriving at precise data collection when the outbreak has reached every country.

The book starts with an introductory chapter on COVID-19. The second chapter informs the reader about the basic principles of infection, epidemics and pandemics. It includes information about the novel coronavirus, epidemiologic triad, chains of infection and the conversion of an outbreak into a pandemic. The third chapter covers the history of pandemics that have caused havoc in the recent past. This chapter serves as a reminder that the world failed to learn from, or rather swept under the rug, the historical facts of great human tragedy. It includes information about the deadly Spanish flu, SARS, MERS, swine flu and Ebola. The fourth chapter informs the reader about the origin of the COVID-19 pandemic in China and how it rapidly swept across the globe, causing immense misery. It focuses on how the epicentre of the pandemic shifted from China to Europe and to the United States and how these countries dealt with it to contain it. The fifth chapter covers the spread of the pandemic in those areas of the world which have lately been badly affected, like Russia, India, Brazil, Iran and Turkey. The epicentre of a pandemic has a high probability of shifting to these areas of the world.

The sixth chapter covers the features of COVID-19. It includes the case definition provided by the WHO and clinical features of and treatments for COVID-19. The seventh chapter deals with the impact of the COVID-19 pandemic: the toll on mental health, disruptions to social life and economic devastation. The eighth chapter includes all the methods for containing the ongoing pandemic. Social distancing, self-isolation, hygienic measures, quarantine, lockdown of cities, the suspension of international travel and border closure are discussed here. Measures for personal protection to prevent infection are adequately discussed. The last chapter covers emerging global issues like vaccine development, pandemic warnings, preparedness and strengthening public health infrastructure. The afterword written at the end provides information related to the COVID-19 pandemic at the time of going to press.

The book relies heavily on information provided by the WHO, the Centers for Disease Control and Prevention (CDC), various highly reputed medical journals and several reputed international media houses. We hesitate to name some for fear of omitting others. We will be able to repay some of their debts if our readers further read about their works and become enlightened. We hope the book will be able to present a clear picture of the COVID-19 pandemic and remove any fear or doubt from their mind. With this singular aim, we submit the book to our readers and authorities in this field for comments, review and constructive criticism.

We express our heartfelt thanks to our family members and friends who helped and encouraged us to complete this task quickly. We acknowledge the contribution of Dr Shashank Shekhar Sinha of Taylor & Francis, India, who helped us to convert the book concept into a reality. We especially thank Ms Brinda Sen and Ms Antara Ray Chaudhary of the editorial team for their untiring effort in realizing this dream project.

Dr Tapas Kumar Koley
Ms Monika Dhole
New Delhi, India
May 2020

PREFACE TO THE SECOND EDITION

The COVID-19 pandemic is not yet showing signs of stopping in its ongoing march of death and destruction throughout the world. It has taken a heavy toll in the last two years. We have been forced to live with the coronavirus for the last two years in an atmosphere of panic, despair and uncertainty. Humanity is trying its best to fight against this virus which is changing its structure frequently and attacking with ever greater force while undermining all our efforts to neutralize it. The WHO, CDC and many other international institutes of repute have carried out extensive research to understand the virus better and devise strategies to minimize the impact of the coronavirus.

The second edition of the book *The COVID-19 Pandemic: The Deadly Coronavirus Outbreak* is written to bring forward all the important information that research undertaken throughout the world has revealed about the ever-changing nature of the virus, its varied clinical manifestations, treatment methods and effects of the vaccines that have been developed at a warp speed to curtail the death and debility caused by coronavirus.

The book contains three new chapters. The first new chapter covers all that has been discovered about the structural changes in the virus that have occurred over the years resulting in formation of several variants that have caused wave after wave of infection leading to a rising number of cases and an ever-increasing death toll. This chapter includes information about the coronavirus variants Alpha, Beta, Gamma, Delta and Omicron and its sublineages. Their origin, mutation, disease-causing potential, response to treatment and efficacy of vaccines against them are discussed in detail.

The second new chapter covers the global spread of COVID-19. This chapter includes the crest and trough of every wave of infection throughout the world. The effect of the variants and their role in the genesis of these waves in various countries of the world are included in this chapter. The death and destruction brought by the coronavirus have been elucidated not only with the help of statistical data from reputed websites of the WHO, Johns Hopkins and the CDC, but also an effort has been made to highlight the emotional trauma, sense of hopelessness and panic created by these waves amongst humanity.

The third new chapter contains information that has come forward in the last two years about the wide spectrum of clinical manifestations of COVID-19, especially amongst children, pregnant women and its lingering effect in the form of Long Covid. The efficacy of new medicines, especially monoclonal antibodies and antiviral medications developed over the last two years, has been highlighted in this chapter. The various aspects of COVID-19 vaccines including their efficacy, side effects, additional doses, booster doses, indications and contraindications and hesitancy of vaccination are also covered in details in this third chapter.

While writing this book reputed medical journals have been referred to and all efforts have been made to keep the medical jargon minimized to make it an easy read for everybody. We hope the book will inform its readers in detail about the latest various aspects of COVID-19 and make them understand the disease better and help them to live with COVID-19, which does not seem to be going away anytime soon.

We take this opportunity to thank our family members and friends who encouraged us and helped us in many ways to write this second edition. At the same time we express our gratitude to the editorial team of Routledge without whom this edition would not have seen the light of the day.

Dr Tapas Kumar Koley
Ms Monika Dhole
New Delhi, India
May 2022

1

INTRODUCTION

The world is fighting a global war against a newly discovered strain of coronavirus. The new strain of coronavirus, named SARS-CoV-2 by International Committee on Taxonomy of Viruses, was first reported in Wuhan, Hubei province, China, in December 2019 and has now rapidly spread across almost all the countries of the world. The disease caused by SARS-CoV-2 infection was named COVID-19 (short version of coronavirus disease 2019, previously known as 2019 novel coronavirus) by the WHO.[1] The virus has genetic similarity to the SARS-CoV virus which was responsible for the severe acute respiratory syndrome (SARS) epidemic in 2002–2003.

Coronavirus infection not only has caused disease and death but has also affected almost every aspect of human life. It has resulted in the disruption of daily lives with cities and countries under lockdown, with many international sporting events, social events, marriages and other ceremonies being postponed or cancelled, triggering a global economic crisis. Almost every industry has been affected throughout the world. Stock markets have crashed. Airlines, the hospitality industry and the travel and tourism sector have been crippled. It has also led to widespread job losses across sectors.

The world was not ready to face such a crisis of such proportions. Lessons from the past were not heeded, and the warning signs were not taken seriously enough. Healthcare workers, first responders, essential service providers and others on the frontlines have taken a direct hit. They are trying their best to fight against the deadly virus with little protective gear and are falling prey to the virus at an alarming rate.

In the past, epidemics were associated with overcrowding, lack of health and hygiene and poor sanitation facilities and yet did not spread at this scale globally. Travelling by ship and later by air enabled a local outbreak to spread globally and take the shape of a pandemic. In 1968, influenza became the first pandemic to

DOI: 10.4324/9781003345091-1

spread by air travel.[2] It was followed by an epidemic of acute haemorrhagic conjunctivitis that spread between international airports.[3] This was the dawn of the modern era of epidemics, where infections in one country could easily spread to another in a short time. Human movement, environmental factors, rapid population rise, inadequate sanitation facilities, poverty and the impact of human activities on the ecosystem have vastly increased the possibility of infectious diseases and the development of pandemics in recent times. It has reached a point where it will suffice to point out that we human beings are now our own enemies.

According to the WHO, by May 15, 2020, there were 4,338,658 confirmed cases of COVID-19 and the death toll had reached 297,119 people globally (Figure 1.1 and Figure 1.2).[4] The progression of the global count of confirmed cases and the mounting death toll are depicted in Figure 1.3 and Figure 1.4. However, there is variation in the data collected by reputed institutions like Johns Hopkins University and the WHO. The reasons are many. It depends on how a case of COVID-19 is defined and on the method for collecting data. Efforts to tabulate the data became more difficult because test kits were not ready when the pandemic started while many people were affected, and the world suffered casualties even before the right diagnosis could be made. True numbers probably will never be fully known; however, data do not always tell the story of misery, death, despair and horror inflicted by a deadly foe.

The WHO declared the coronavirus outbreak a public health emergency of international concern on January 30, 2020, and a 'pandemic' on March 11, 2020.[5] Declaration of pandemic and widespread death all across the world and especially in developed countries of the West has resulted in widespread fear and panic. The absence of medicine and vaccine for the treatment and prevention of coronavirus infection has added fuel to fire. The medical community had very little knowledge at the beginning about this faceless enemy.

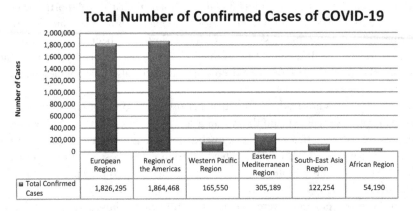

FIGURE 1.1   Total Number of Confirmed Cases of COVID-19 Worldwide by May 15, 2020

Source: WHO.

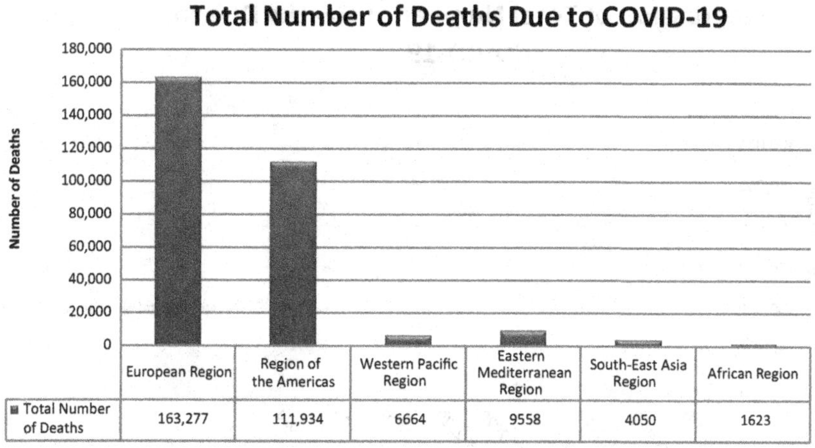

Total Number of Deaths Due to COVID-19
Number of Deaths
180,000
160,000
140,000
120,000
100,000
80,000
60,000
40,000
20,000
0
European Region
Region of the Americas
Western Pacific Region
Eastern Mediterranean Region
South-East Asia Region
African Region
Total Number of Deaths
163,277
111,934
6664
9558
4050
1623

FIGURE 1.2 Total Number of Deaths Due to COVID-19 Worldwide by May 15, 2020

Source: WHO.

Progression of Total Number of Confirmed Cases of COVID-19
Number of Cases
5,000,000
4,500,000
4,000,000
3,500,000
3,000,000
2,500,000
2,000,000
1,500,000
1,000,000
500,000
0
January 24, 2020
February 7, 2020
February 21, 2020
March 6, 2020
March 20, 2020
April 3, 2020
April 17, 2020
May 1, 2020
May 15, 2020
Total Confirmed Cases

FIGURE 1.3 Progression of Total Number of Confirmed Cases of COVID-19 Worldwide

Source: WHO.

Many questions have been raised regarding the role of China,[6] where the outbreak started, and the role of the WHO in the mitigation of the COVID-19 pandemic. China is facing a global backlash over the mishandling of the coronavirus outbreak. Australia has called for an inquiry into the origin of the coronavirus infection. US President Donald Trump has blamed China for the contagion and

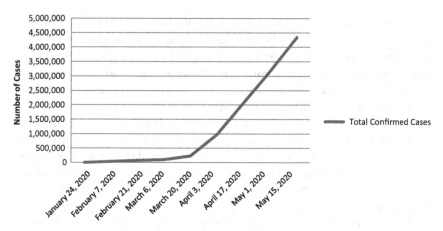

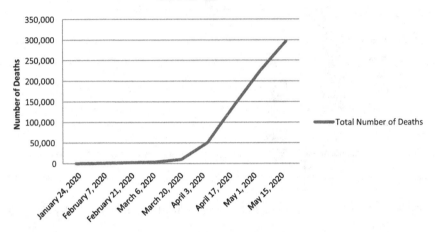

FIGURE 1.4 Progression of Total Number of Deaths Due to COVID-19 Worldwide

Source: WHO.

is seeking to punish it.[7] At the same time, the WHO is facing mounting pressure to answer critics worldwide who say Chinese influence has hindered its response to the coronavirus pandemic.[8] President Trump halted funding to the WHO over its handling of the coronavirus pandemic.[9] However, this is not the time to play the blame game; rather, concentrated efforts to minimize the damage and halt the spread of the pandemic should take priority.

Against this backdrop, *COVID-19 Pandemic: The Deadly Coronavirus Outbreak* is written to filter out the facts from the fiction and enable everybody to have a true picture of this unprecedented human crisis. The book is divided into nine chapters to cover all the important information on COVID-19. It starts with this introductory chapter, wherein a brief introduction to COVID-19 is covered. The second chapter covers the basic principles of an infectious disease and how an infection like COVID-19 turns into a pandemic. Basic information on epidemics, pandemics, immunity, pathogens, chains of infection and the transmission of infection are provided. Readers will develop an idea about case fatality rate, incubation periods and reproductive number (R_0), which will help them to understand the pandemic as they read the following chapters.

The third chapter provides a glimpse into the various deadly pandemics that have happened in the past, including the Spanish flu, SARS, MERS, swine flu, Ebola and Zika. How these pandemics started and affected humankind is discussed in brief. Its short revisit of the past enables readers to compare past pandemics with COVID-19. Readers will learn about the destructive and disruptive effects of pandemics.

The fourth chapter focuses on the origin and spread of the pandemic and the global response to the pandemic. The infection originated at the Huanan Seafood Wholesale Market in Wuhan, China,[10] and then spread among the residents of Wuhan. Rapidly the infection spread, and casualties started, resulting in the massive mobilization of healthcare resources to control the outbreak. Gradually, it spread to various countries of the world, and later, the WHO declared it a pandemic. This chapter provides a timeline of evolution of this epidemic and how it changed from an epidemic to a pandemic in a short period. It focuses on how the epicentre of the disease shifted from China to highly developed countries in Europe and the United States. The evolution of the disease and the number of cases and deaths in various countries, like China, Italy, France, Spain, Germany, the United Kingdom, the United States, Australia and Canada, are discussed.

The fifth chapter covers the spread of the pandemic in Russia, Brazil, Turkey, Iran, India and rest of the world. Russia, Brazil and Turkey rapidly turned out to be among the worst-affected countries of the world. The number of cases has increased lately to a great extent. Similarly, of late, the number of cases has also increased in India. The evolution of epidemic in these countries is discussed, along with the response of the government to the pandemic. Conditions in Africa are also included in this chapter.

The sixth chapter covers the signs and symptoms of COVID-19. How to identify a case of COVID-19 is discussed. How COVID-19 starts with fever and cough and rapidly evolves into pneumonia and respiratory failure and death in a short period is discussed. The reader will learn about the common as well as uncommon symptoms of the disease. Steps to be taken if one falls sick, infection prevention protocol, laboratory investigations for making a confirmed diagnosis and health tips are discussed in this chapter. Medical information in the chapter is provided in plain language, avoiding medical jargon.

The focus of the seventh chapter is on mental trauma, socioeconomic disruption and the economic crisis caused by COVID-19. The impact of the pandemic on various aspects like trade and commerce, disruptions and postponements of sports events, psychological effects on people, rumours, violations of government orders and disruptions to daily life are discussed. The economic crisis triggered by the pandemic is covered in this chapter. Efforts by various governments and financial institutions to support the economy are also discussed here.

The eighth chapter deals with containing the coronavirus pandemic. Coronavirus infection has no effective medicine or vaccine, and hence, the prevention of infection is the most important aspect in containing this devastating pandemic. Steps to be taken for preventing infection and for containing the pandemic are discussed from personal, national and global perspectives. WHO and CDC guidelines on important issues like use of face masks, hand washing, alcohol-based sanitizer, self-isolation and social distancing are discussed in great detail. Emphasis has been placed on locking down cities and on quarantine, and they have played a significant role in the containment of this unprecedented pandemic. Important aspects like the discontinuation of visas, cancellations of airplanes and trains and other measures to control the movement of people that help in the containment of a pandemic have been stressed. All efforts by health authorities to flatten the epidemic curve so that health facilities are not overwhelmed are dealt with in this chapter.

The ninth chapter covers the global challenges that have emerged as a result of the COVID-19 pandemic. Important vital issues like efforts of vaccine development, pandemic warnings and the outcome of ignoring the warnings and importance of strengthening the public health system are covered in detail. These critical global challenges are not only important in the current pandemic but will be extremely beneficial in the future as well. Afterword at the end of the book contains latest information about COVID-19 pandemic at the time of going to press.

COVID-19 reminds us of the possibility of emerging and re-emerging infectious agents and the need for constant surveillance and for prompt diagnoses and treatments to understand the disease-causing mechanism of these agents and to develop effective countermeasures. Robust and effective countermeasures are the need of the hour, to save humanity from this viral Armageddon which has the capacity to wipe out human existence. *COVID-19 Pandemic: The Deadly Coronavirus Outbreak* is a humble effort in this direction.

Notes

1 Naming the coronavirus disease (COVID-19) and the virus that causes it, WHO, available at www.who.int/emergencies/diseases/novel-coronavirus-2019/technical-guidance/naming-the-coronavirus-disease-(covid-2019)-and-the-virus-that-causes-it, accessed on March 25, 2020
2 Longini I, Fine P, Thacker ST, Predicting the global spread of new infectious agents, *Am J Epidemiol*, 1986; 123: 383–391
3 Morens DM, Daszak P, Taubenberger JK, Escaping Pandora's Box – Another novel coronavirus, *NEJM*, available at www.nejm.org/doi/full/10.1056/NEJMp2002106, accessed on March 25, 2020
4 Coronavirus disease (COVID-19) situation report – 116, WHO, available at www.who.int/docs/default-source/coronaviruse/situation-reports/20200515-covid-19-sitrep-116.pdf?sfvrsn=8dd60956_2, accessed on May 16, 2020
5 WHO characterizes COVID-19 as a pandemic, available at www.who.int/emergencies/diseases/novel-coronavirus-2019/events-as-they-happen, accessed on March 25, 2020
6 Coronavirus: US and China trade conspiracy theories, *BBC*, available at www.bbc.com/news/world-52224331, accessed on May 9, 2020
7 COVID-19: Global backlash builds against China over coronavirus, *The Times of India*, available at https://timesofindia.indiatimes.com/world/china/covid-19-global-backlash-builds-against-china-over-coronavirus/articleshow/75531645.cms, accessed on May 17, 2020
8 Cooper S, Why is the World Health Organization accused of mishandling the coronavirus pandemic? *Global News*, available at https://globalnews.ca/news/6826415/world-health-organization-accused-of-mishandling-coronavirus-pandemic-covid-19/, accessed on May 16, 2020
9 Donald Trump stops US funding to WHO, blames it for COVID-19 spread, *The Wire*, available at https://thewire.in/world/donald-trump-us-who-funding-covid-19, accessed on May 9, 2020
10 Maron DF, 'Wet markets' likely launched the coronavirus. Here's what you need to know, *National Geographic*, available at www.nationalgeographic.com/animals/2020/04/coronavirus-linked-to-chinese-wet-markets/, accessed on May 10, 2020

2

PANDEMIC

The basic principles

With the onset of coronavirus infection in different parts of the world, many of us have been exposed to terminology related to the disease, such as pandemic, epidemic, endemic, incubation period, case fatality rate and many more. Although we have some idea about these terms, we may not have a complete understanding of these words; neither do we know their implications for a deadly infectious disease like COVID-19.

To understand the evolution and spread of coronavirus infection, it is important to understand the meaning of these words, including the difference between 'epidemic' and 'pandemic'.

Basic terminology

Outbreak

When an infectious disease remains localized to a small geographic area, it is called an outbreak. An outbreak carries more or less the same significance as an epidemic.

Epidemic

A relatively known term to all of us, an epidemic means an increase, often sudden, in the number of cases of an infectious disease above what is normally expected in a population of a given area in a specific time period.

Pandemic

An epidemic that spreads over several countries or continents, usually affecting a large number of people, is called a pandemic.

DOI: 10.4324/9781003345091-2

Coronavirus infection may be called an outbreak when it was localized in the city of Wuhan, China. It turned into an epidemic when it spread rapidly and involved a large majority of people in several cities of China. However, it took the form of a pandemic when it crossed the borders of China and spread in several other countries and continents and affected a huge population.

Sporadic, endemic and hyperendemic

In terms of frequency of occurrence, a disease or an infection can be called sporadic, endemic and hyperendemic. A disease that occurs infrequently and irregularly is called sporadic, whereas a disease or any infectious agent that remains present in a population within a geographic area constantly is called endemic, and it becomes hyperendemic if the occurrence of the disease becomes high and persistent in a population.

Incubation period

There is a time interval between the exposure of a susceptible host (a person or animal) to an infectious agent (virus, bacteria, protozoa, etc.) and the appearance of the first signs or clinical symptoms of the disease in that host. This time period is referred to as the incubation period of a disease.

According to the WHO, the incubation period for COVID-19 ranges from 1 to 14 days, most commonly around five days.[1] This means that after a healthy person comes into contact with a person who has a coronavirus infection, they are likely to develop infection within 14 days of contact.

Case fatality rate

The case fatality rate is another important aspect of any epidemic or pandemic. It is the proportion of deaths in relation to the number of reported cases of a disease (i.e. the number of affected people by a disease) within a specified time. The case fatality rate of the coronavirus pandemic has been found to vary by age, sex and presence or absence of coexisting diseases like diabetes, hypertension, respiratory diseases, etc. It also has regional variations. The overall fatality rate of people with confirmed COVID-19 in the Italian population, based on data up to March 17, 2020, was 7.2% (1625 deaths/22,512 cases).[2] The overall case fatality rate in Italy (7.2%) is substantially higher than that in China (2.3%). When analyzed by age group, the case fatality rate in Italy and China appear similar for age groups 0 to 69 years, but rates are higher in Italy among individuals aged 70 years or older and in particular among those aged 80 years and above.[3] However, the exact case fatality rate is difficult to estimate because not all countries are testing to detect the infection. Besides this, the pandemic is still evolving, and the ultimate result may be far different from what it is at present.

Basic reproduction number (R_0)

Another key concept in understanding the coronavirus pandemic is the basic reproduction number, R_0,[4] which was originally named the basic reproductive ratio or rate.[5]

In terms of the reproductive number, there are three possible situations:

1   R_0 = 1 indicates that the number of cases is stable and the disease is endemic.
2   R_0 > 1 indicates that the number of cases is increasing and that it will ultimately turn into an epidemic.
3   R_0 < 1 indicates that the number of cases is decreasing and that the epidemic will come under control soon (Figure 2.1).

In China, the R_0 for COVID-19 is 2–2.5, which means that on average, each patient transmits the infection to an additional 2.2 individuals.[6] In Italy, the basic reproduction number ranges from 2·76 to 3·25.[7] These data indicate that coronavirus infection is a highly contagious disease and that the epidemic may take some time to settle down (Figure 2.2).

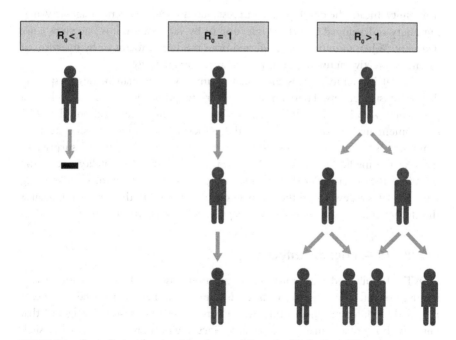

FIGURE 2.1   Basic Reproduction Number and Spread of Infectious Disease
Source: Author.

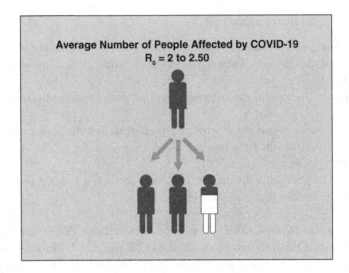

FIGURE 2.2   Spread of COVID-19 on the Basis of R_0

Source: Author.

Immunity

'Immunity' means the development of resistance against an infectious agent. When immunity is developed by natural infection or by vaccination, it is known as *active immunity*. When immunity is transferred from mother to foetus or by injection of specific protective antibodies, it is known as *passive immunity*.

'Natural immunity' means the innate immunity of human beings to specific disease-causing agents. Herd immunity refers to proportion of individuals in a population who have immunity against a specific disease-causing agent. The UK government initially did not prevent the gathering of people, to generate herd immunity as a strategy to fight against coronavirus epidemic, to the surprise of many in the medical community. However, realizing the high fatality rate of the infection, the government shelved this strategy.[8] Because the coronavirus-causing COVID-19 is a new agent, there is negligible immunity in the population against this disease, and almost everyone is susceptible to coronavirus infection.

COVID-19 – a deadly infection

COVID-19 falls under the category of infectious diseases. Hence, a basic under-standing of infectious disease, in brief, is appropriate here. An infectious disease is caused by different types of microscopic organisms (disease-causing agents) that enter the body of a human or animal and affect one or more vital body parts, such as the lungs, intestines or urinary tract. These organisms, which can be present any-where around us, can enter the body through nose, mouth or skin and can multiply thereafter, which may or may not lead to an illness or disease.

Epidemiologic triad

To understand the development of an infectious disease such as COVID-19 in a human body, a traditional model is used all over the world: epidemiologic triad or triangle (Figure 2.3). This triad consists of first a disease-causing agent, second a susceptible host (human being) and third an environment that brings the agent and the host into proximity. A disease occurs when the agent interacts (comes close) with the susceptible host in an environment that supports or allows for transmission of the agent into the body of the host.

Agent

The agents or the disease-causing micro-organisms are also called pathogens. These include bacteria, viruses, fungi and protozoa. The development of a disease depends on several factors related to these agents, like their pathogenicity (ability to cause a disease) and dose (quantity). The mere presence of micro-organisms in the environment is not sufficient to cause a disease. Usually, the greater the number of pathogens in the body, the more severe the disease, and the same has been observed in COVID-19.

Bacteria

Bacteria are microscopic organisms that are capable of producing many types of diseases. Cholera, plague and typhoid are some of the important diseases caused by bacteria. Bacteria have the ability to reproduce by themselves inside a human body,

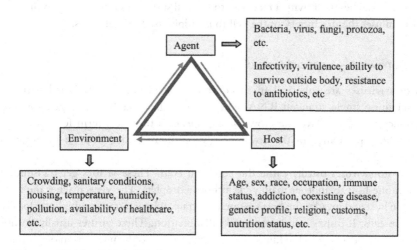

FIGURE 2.3 Epidemiologic Triad

Source: Author.

thereby increasing the severity of a disease. Some of them are round in shape, and some of them are rod shaped. In some, there is a hair-like structure over the body.

Fungi

Fungi are another type of microscopic pathogen. Skin infections and pneumonia are some of the important diseases caused by them. Fungi are often compared with plants, because they are made up of many cells, unlike bacteria or viruses. However, they are not capable of producing food of their own. So, they depend on human beings and plants for their survival. They typically cause infection when the host has low immunity.

Protozoa

Protozoa are unicellular organisms found worldwide in most habitats. Most species are free living, have relatively complex internal structure and carry out complex metabolic activities. Some species are considered commensals, i.e. normally not harmful, whereas others are pathogens and usually produce disease. Infections caused by them range from asymptomatic to life-threatening conditions.[9] Malaria is an important disease caused by protozoa.

Viruses

Viruses are also microscopic organisms, much smaller in size than bacteria. They cause a lot of diseases, like measles, chicken pox, the common cold, etc. An important difference between a virus and a bacterium is that a virus is not capable of multiplying by itself, although it has its own genetic material. Once it enters a human body, it attaches itself, with a receptor, to the cell surface and gains entry to the cell. It then uses the machinery of the cell to multiply itself several times.

Coronavirus – a special type of virus

Coronaviruses are spherical in shape and are made up of enveloped particles containing single-stranded RNA. The envelope bears club-shaped glycoprotein projections. They have the crown-like (*coronam* is the Latin term for crown) or halo-like appearance of the envelope glycoproteins under electron microscope (Figure 2.4).[10]

Coronaviruses usually cause the common cold. These viruses get transmitted from one host to another mainly via airborne droplets. The virus reaches the nose of the host and invades the host's respiratory tract, where it starts multiplying locally in the cells. It causes cell damage and inflammation. These viruses usually do not cause much damage to the lungs. The infection leads to the production of antibodies which ultimately clear the virus from the body, and the infection subsides. The antibody may last for a year or two to provide immunity for future infections.

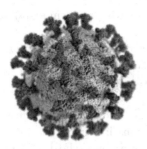

FIGURE 2.4 Diagram of Coronavirus

Source: CDC, PHIL.

Photo credit: Alissa Eckert, MS, and Dan Higgins, MAMS.

Maximum infection occurs during winter season and may take the form of a local epidemic. It is difficult to diagnose the infection and differentiate it from other causes of the common cold. Treatment is symptomatic, as there is no specific treatment.

Coronavirus has gained notoriety in this century because it has caused an epidemic of massive proportion in recent past. For reasons yet to be explained, these viruses can cross species barriers and have caused severe diseases such as SARS and MERS. These viruses have probably originated from bats and then moved into other mammalian hosts – the Himalayan palm civet for SARS-CoV and the dromedary camel for MERS-CoV – before jumping to humans.[11]

The current pandemic caused by SARS-CoV-2 is a new strain of coronavirus which seems to be very contagious (due to mutation) and has quickly spread globally. It belongs to the betaCoVs category and has a diameter of approximately 60–140 nanometres. Like other coronaviruses, it is susceptible to heat, ultraviolet rays, ethanol, 75% ether and chlorine-containing disinfectants. The isolates from the Wuhan strain shares a 89% nucleotide identity with bat SARS-like-CoVZXC21 and 82% with that of human SARS-CoV,[12] and hence, it has been named SARS-CoV-2.

The SARS-CoV-2 spike protein binds with the human cell surface of the angiotensin-converting enzyme (ACE2) receptor and thus gains entry to the cell. ACE2 is expressed by the epithelial cells of the lungs, intestine, kidney and blood vessels.[13] After entering cells, it starts causing damage to the cells, leading to inflammation and extensive tissue damage. In the majority of cases, it causes mild symptoms, and the patient may even be asymptomatic, but in certain cases, it causes severe damage to lungs, leading to death.

Host

In the epidemiologic triad, the host is a human being or an animal that either suffers or harbours the disease-causing agents. The host may become sick or may remain asymptomatic (not showing any symptoms) and carry the agent silently, spreading it in the community.

People are not equally susceptible to the disease. There are several factors intrinsic to the host, called risk factors, that determine the susceptibility to the disease. These important risk factors are age, sex, race, immune status, smoking and drinking habits, drug addiction, occupation, nutritional status, genetic composition, coexistence with other diseases and psychological makeup.

Reports suggest that coronavirus infection is more likely to affect old male patients. The median age of patients who have COVID-19 ranges from 49 to 59 years.[14] Infection in children is being reported much less commonly than among adults, and all cases so far have been in family clusters or in children who have a history of close contact with an infected patient.[15]

Most of the COVID-19 patients have been found to have an underlying comorbid (co-occurring) health condition like diabetes, hypertension, respiratory disease or kidney dysfunction. Clinical and epidemiological data from the Chinese CDC (Centers for Disease Control and Prevention)[16] regarding 72,314 case records reported in the *Journal of the American Medical Association* has revealed that the fatal cases were primarily elderly patients, in particular those aged over 80 (about 15%) and those aged 70 to 79 (8.0%). About half (49.0%) of the serious patients who had pre-existing comorbid conditions, such as cardiovascular disease, diabetes, chronic respiratory disease and cancer, died. [17]

Similarly, data from the United States has revealed that between February 12 and March 16, 2020, 4226 COVID-19 cases were reported in the United States. Among them, 31% of cases, 45% of hospitalizations, 53% of intensive care unit (ICU) admissions and 80% of deaths occurred among adults aged 65 and over with the highest percentage of severe outcomes among adults aged 85 and over.[18] These data are similar to data obtained from the Chinese CDC.

However, this does not mean that young people are not suffering or are not being hospitalized. The same report of the US CDC revealed that among the hospitalized patients, 20% were aged 20–44, and among all the patients that needed ICU care, 12% were in the age group 20–44.

According to the *Washington Post* in an article published on March 19, 2020, French health ministry official Jérome Salomon said that half of the 300 to 400 coronavirus patients treated in intensive care units in Paris were younger than 65, and according to numbers presented at a seminar of intensive care specialists, half the ICU patients in the Netherlands were younger than 50.[19]

Reports from Italy have revealed that the majority of the affected patients are from the older age group (age more than 60 years) and maximum death has also occurred in the older age group (individuals aged 70 years or older, and in particular among those aged 80 years or older).[20] Italy has a high proportion of older patients with confirmed COVID-19 infection, and the older population in Italy (approximately 23% of the Italian population was aged 65 years or older) may partly explain the high mortality rate in this age group.

Environment

Environment comprises all the external factors that affect the host, the agent and their interactions. It includes physical factors such as temperature, humidity and geology; socioeconomic factors such as poverty, pollution, sanitation, crowding, availability of health services and psychological makeup; and biological factors such as insects that transmit the agent to cause a disease.

To diagnose and understand a disease, the doctor must consider the patient in the context of the population to which they belong. Infectious diseases often do not occur in isolation but rather occur in a group of people who are exposed to the agent from a single source. Early cases identified in Wuhan were believed to have acquired infection from an animal source, because many reported visiting or working in the Huanan Wholesale Seafood Market.[21]

Another environmental factor that is crucial in the current pandemic is the history of travel. People with history of travel to China, Italy, France, Iran and other coronavirus-affected countries were screened and many put into quarantine in India and many other countries to control the spread of infection by early detection and isolation.[22]

A combination of factors like overcrowding, altered human behaviour, environmental changes, lack of adequate healthcare facilities and air travel has led to the development of a human-dominated ecosystem that provides favourable conditions for the emergence of dangerous microscopic organisms, especially rapidly mutating RNA viruses, which wreak havoc on people and pose existential threats to humankind.

Prevention

To prevent or control an infectious disease, physicians try to understand the complex interaction between the agent, host and environmental factors and devise strategies to break the interaction between these three limbs of the triangle. For instance, during this ongoing coronavirus pandemic, social isolation and staying at home have been used as strategies to control the onslaught of infection, by breaking the chain of transmission of coronavirus from one person to another (Figure 2.5). The preventive methods are discussed in a separate chapter.

Infection to pandemic

An infectious disease like COVID-19 takes the form of an epidemic or a pandemic when an agent and the susceptible hosts are present in an environment in adequate numbers and the agent is effectively and rapidly transmitted to the susceptible hosts. More specifically, an epidemic may result from the following:[23]

- Entry of a new agent into an environment where it was absent before
- A recent and sudden increase in the agent's level or virulence (capacity to produce disease)

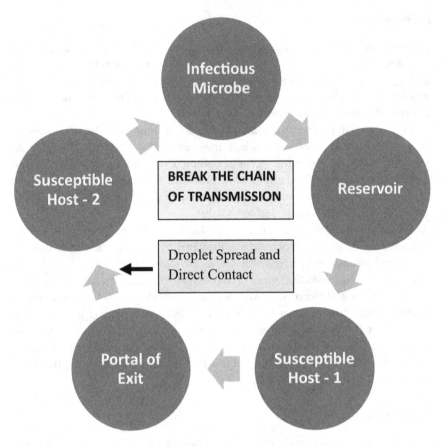

FIGURE 2.5 Break the Chain of Transmission

Source: Author.

- An enhanced mode of transmission so that more susceptible people are exposed
- A change in the susceptibility of the host to the agent
- Factors that increase host exposure or involve introduction through new portals of entry.[24]

Once an infectious agent, such as influenza, establishes itself during an epidemic, it continues to infect more and more people, and the disease spreads from person to person from one place to another and ultimately across the borders of the country till it takes the shape of a pandemic. However, it gradually settles into a cyclic mode wherein susceptible people are infected who become immune to any further attack. As the number of susceptible people reduce, the virus moves on only to return when the next generation of susceptible population consisting of those who are born after the last epidemic appear in the environment.[25] The other reasons for a cyclical trend of an epidemic are the emergence of a new group of susceptible population, due to the decline in protective immunity, and the removal of immune individuals, due to death.

Transmission of infection

Transmission of infection is central to the causation of a disease, and the transmission modes of pathogens are complex and diverse. A thorough understanding of the mode of transmission of an infectious agent is necessary to understand the evolution of a disease, its transformation into an epidemic or pandemic, which in turn enables us to determine the strategies to control and contain the epidemic. A chain of infection is established when an infectious agent comes out from its reservoir or host through an exit point (portal of exit), gets carried by some mode of transmission and enters the body of a susceptible host through an appropriate entry point (portal of entry). Establishing a chain of infection is necessary for an infection to occur and maintaining this chain and further propagating it in a brisk manner causes an epidemic to develop.

Reservoir

The reservoir of a pathogen is the habitat or home where it finds suitable conditions that permit it to reside, grow and multiply for a long period of time. The reservoir may be living or non-living and may include humans, animals or the environment. Common infectious diseases like measles and mumps have human reservoirs.

A carrier is a person who harbours the pathogen and can transmit the pathogen to others but does not have the illness. These asymptomatic healthy carriers are important sources of the transmission of pathogens to another susceptible host. During the current pandemic, coronavirus has spread via these types off healthy carriers.[26] Young people get infected but do not show any signs of illness. Neither are they suspected to be suffering, nor are they tested to see whether they are carriers. They move around in society and go on infecting others. The WHO report highlights that out of some 3664 cases which were reported outside of China, 92 were detected while apparently asymptomatic.[27]

Similarly, many pathogens reside in animals, and they occasionally jump from animals to humans. HIV, Ebola and SARS evolved from animals to cause widespread disease in recent times. The ecological reservoir of SARS-CoV-2 is thought to be bats, from which the disease has spread to human beings.[28]

Portal of exit

The path taken by the pathogen to exit from the body of host is called the portal of exit. For example, tuberculosis-causing bacteria exit via the respiratory tract, and *Vibrio cholerae* (bacteria that cause cholera) exits through faeces. The portal of exit for SARS-CoV-2 is mainly the nose and the mouth. Further research is required to identify all the possible portals of exit.

Mode of transmission

There are several modes of transmission of a pathogen from reservoir to a susceptible host. This susceptible host further transmits this pathogen to other people and propagates the chain of infection.

Direct contact and droplet spread

Direct contact and *droplet spread* are two main modes of transmission of coronavirus infection. Droplet spread refers to the spray of pathogens with relatively large, short-range aerosols produced by sneezing, coughing or even talking. Through droplet spread, the pathogen is transmitted by direct spray over a few feet, before the droplets fall to the ground.

For the spread of SARS-CoV-2, close contact between two people (one is infected, the other healthy) is necessary. Here the pathogen is transmitted by physical contact between two individuals through actions such as touching, kissing, shaking hands or hugging. The virus can also spread directly from person to person when a COVID-19 patient coughs or exhales, producing droplets that reach the nose, mouth or eyes of another person. Alternatively, as the droplets are too heavy to be airborne, they land on objects and surfaces surrounding the person. Other people become infected with COVID-19 by touching these contaminated objects or surfaces, and thereafter touching their eyes, nose or mouth.[29] Therefore, COVID-19 has been found to spread mostly among family members, healthcare professionals who are treating such patients and other people who come into contact with an infected person.

Airborne transmission

Airborne transmission occurs when pathogens are carried by dust or droplet nuclei which remain suspended in air. Airborne transmission of COVID-19 has not yet been confirmed by the WHO. However, research is going on to study whether SARS-CoV-2 is airborne. Given the many similarities between the two SARS viruses and the evidence on virus transport in general, it is highly likely that the SARS-CoV-2 virus also spreads by air.[30,31]

Vehicle-borne transmission

Food, water, blood and fomites (inanimate objects like handkerchiefs, masks, towels and surgical instruments) can transmit pathogens. According to the US CDC, the novel coronavirus has not been detected in drinking water. Although it has been detected in the faeces of some patients who have COVID-19, there have been no reports of faecal–oral transmission of COVID-19.[32] However, viable SARS-CoV-2 in the stool of COVID-19 patients has been detected and reported,[33] and virus RNA has been found in sewage,[34] raising the possibility of faecal–oral transmission.[35]

Available evidence indicates that COVID-19 is transmitted to close contacts by fomites.[36] Towels, handkerchiefs, clothes and masks can be sources of infection to family members and healthcare workers. Hence, the proper disposal of used masks, gloves and gowns is crucial in preventing the spread of the virus.

Vector-borne transmission

Vectors such as mosquitoes, flies, fleas and ticks often carry pathogen and transmit disease. The transmission of malaria by mosquitoes is a well-known example. To date, there is no evidence of vector-borne transmission of COVID-19.

Portal of entry

The portal of entry means the site and the manner in which a pathogen enters the body of a susceptible host. The respiratory tract is the portal of entry of other coronavirus-like respiratory pathogens.

Host

A susceptible host is the final link in the chain of infection. Susceptibility depends on genetic factors, nutritional factors, specific immunity and many unspecific factors that affect an individual's ability to resist. Since SARS-CoV-2 is a new strain of coronavirus, there is hardly any immunity against it in the general population. As a result, the infection is spreading fast and wide and has turned into a pandemic.

Phases of a pandemic

In the context of the COVID-19 pandemic, a common term used in the media is 'phases of pandemics'. The WHO developed various phases of pandemics (Figure 2.6) in 1999 and revised them in 2005 in context of an influenza pandemic, to help countries in pandemic preparedness and response planning. They help to assess the spread of a pandemic and to implement measures for mitigation. The six phases are as follows:[37]

Phase 1 – No viruses circulating among animals have been reported to cause infections in humans.

Phase 2 – An animal influenza virus circulating among domesticated or wild animals is known to have caused infection in humans. Hence, it is considered a potential pandemic threat.

Phase 3 – An animal or human-animal influenza reassortant virus has caused sporadic cases or small clusters of disease in people. However, it has not resulted in human-to-human transmission sufficient to sustain community-level outbreaks.

Phase 4 – Human-to-human transmission of an animal or human-animal influenza reassortant virus able to cause community-level outbreaks has been verified.

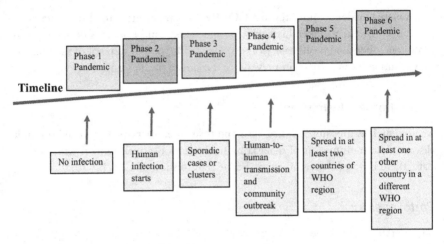

FIGURE 2.6 The Six Phases of Pandemics

Source: Author.

Phase 5 – Human-to-human spread of the virus into at least two countries in one
WHO region.
Phase 6 – Community-level outbreaks in at least one other country in a different
WHO region, in addition to the criteria in Phase 5.

In the case of the current COVID-19 pandemic, the mutated coronavirus has
acted as a deadly agent that has been striking susceptible hosts who have been
falling sick rapidly in the absence of any immunity to the disease. The virus has
spread from one person to another by droplet spread and direct contact. The chain
of transmission which was established has resulted in an outbreak that has quickly
turned into an epidemic due to absence of a vaccine and medicine. Global mobility has carried the virus to faraway places from its place of origin, and thus, the
epidemic has rapidly turned into a deadly viral pandemic.

Humanity has been facing pandemics for a long time. Pandemics affect almost
everybody. Only lucky ones escape the pathogens. The Spanish flu pandemic
spread through almost every country in the world about a century ago. The SARS
pandemic spread through the world in the year 2002–2003. All these pandemics
started in some part of the world and due to extreme infectivity of the pathogen rapidly spread from one country to another with devastating consequences.
A study of these pandemics will enable us to understand the spread and consequences of pandemics.

Notes

1 Report of the WHO-China joint mission on coronavirus disease 2019 (COVID-19),
available at www.who.int/docs/default-source/coronaviruse/who-china-joint-mission-
on-covid-19-final-report.pdf, accessed on March 26, 2020

2 Livingston E, Bucher K, Coronavirus disease 2019 (COVID-19) in Italy, *JAMA*, published online March 17, 2020, doi:10.1001/jama.2020.4344

3 Onder G, Rezza G, Brusaferro S, Case-fatality rate and characteristics of patients dying in relation to COVID-19 in Italy, *JAMA*, published online March 23, 2020, available at https://jamanetwork.com/journals/jama/fullarticle/2763667, accessed on March 26, 2020

4 Fraser C, Riley S, Anderson RM, et al., Factors that make and infectious disease controllable, *Proc Nat Acad Sci*, 2004; 101: 6146–6151

5 Scott ME, Smith G, eds. *Parasitic and Infectious Diseases: Epidemiology and Ecology*, San Diego: Academic Press, 1994

6 Cascella M, Rajnik M, Cuomo A, et al., *Features, Evaluation and Treatment Coronavirus (COVID-19)*, available at www.ncbi.nlm.nih.gov/books/NBK554776/, accessed on March 26, 2020

7 Remuzzi A, Remuzz G, COVID-19 and Italy: What next? *Lancet*, published online March 12, 2020, https://doi.org/10.1016/S0140-6736(20)30627-9, accessed on March 26, 2020

8 O'Gardy C, The U.K. backed off on herd immunity: To beat COVID-19, we'll ultimately need it, available at www.nationalgeographic.com/science/2020/03/uk-backed-off-on-herd-immunity-to-beat-coronavirus-we-need-it/, accessed on March 26, 2020

9 Yaeger RG, *Protozoa: Structure, Classification, Growth, and Development*, Medical Microbiology, 4th edition, available at www.ncbi.nlm.nih.gov/books/NBK8325/, accessed on March 26, 2020

10 Tyrrell DAJ, Myint SH, *Coronaviruses*, Medical Microbiology, 4th edition, available at www.ncbi.nlm.nih.gov/books/NBK7782/, accessed on March 26, 2020

11 Cascella et al., *Features, Evaluation and Treatment Coronavirus (COVID-19)*

12 Chan JF, Kok KH, Zhu Z, et al., Genomic characterization of the 2019 novel human-pathogenic coronavirus isolated from a patient with atypical pneumonia after visiting Wuhan, *Emerg Microbes Infect*, 2020; 9(1): 221–236

13 Wan Y, Shang J, Graham R, et al., Receptor recognition by novel coronavirus from Wuhan: An analysis based on decade-long structural studies of SARS, *J Virology*, 2020, published online January 29, doi:10.1128/JVI.00127-20

14 Huang C, Wang Y, Li X, et al., Clinical features of patients infected with 2019 novel coronavirus in Wuhan, China, *Lancet*, 395(10223): 497–506, January 24, 2020; Chen N, Zhou M, Dong X, et al., Epidemiological and clinical characteristics of 99 cases of 2019 novel coronavirus pneumonia in Wuhan, China: A descriptive study, *Lancet*, 395(10223): 507–513, January 30, 2020; Wang D, Hu B, Hu C, et al., Clinical characteristics of 138 hospitalized patients with 2019 novel coronavirus-infected pneumonia in Wuhan, China, *JAMA*, 2020; 323(11): 1161–69, February 7, 2020

15 Chen ZM, Fu JF, Shu Q, et al., Diagnosis and treatment recommendations for pediatric respiratory infection caused by the 2019 novel coronavirus, *World J Pediatr*, 2020 Jun; 16(3): 240–246, February 5, 2020; Shen KL, Yang YH. Diagnosis and treatment of 2019 novel coronavirus infection in children: A pressing issue, *World J Pediatr*, 16: 219–221 (2020), February 5, 2020

16 www.chinacdc.cn/en/ (website address for Chinese CDC)

17 Wu Z, McGoogan JM, Characteristics of and important lessons from the coronavirus disease 2019 (COVID-19) outbreak in China: Summary of a report of 72 314 cases from the Chinese Centers for Disease Control and Prevention, *JAMA*, 2020; 323(13): 1239–1242, February 24, 2020

18 Severe outcomes among patients with coronavirus disease 2019 (COVID-19) – United States, February 12–March 16, 2020, CDC, *Morb Mortal Wkly Report (MMWR)*, March 18, 2020, available at www.cdc.gov/mmwr/volumes/69/wr/mm6912e2.htm, accessed on March 26, 2020

19 Younger adults make up a large percentage of coronavirus hospitalizations in the United States, according to new CDC data, by Ariana Eunjung Cha, available at www.washingtonpost.com/health/2020/03/19/younger-adults-are-large-percent

age-coronavirus-hospitalizations-united-states-according-new-cdc-data/, accessed on March 26, 2020

20 Onder et al., Case-fatality rate and characteristics of patients dying in relation to COVID-19 in Italy

21 Report of the WHO-China joint mission on coronavirus disease 2019 (COVID-19)

22 What we scientists have discovered about how each age group spreads COVID-19, *The Guardian*, available at www.theguardian.com/commentisfree/2020/mar/17/scientists-age-groups-covid-19-workplaces-shops-restaurants, accessed on May 16, 2020

23 *Principles of Epidemiology in Public Health Practice*, 3rd edition, An Introduction to Applied Epidemiology and Biostatistics Introduction to Epidemiology Section 11: Epidemic Disease Occurrence, available at www.cdc.gov/csels/dsepd/ss1978/lesson1/section11.html, accessed on March 26, 2020

24 Kelsey JL, Thompson WD, Evans AS, *Methods in Observational Epidemiology*, New York: Oxford University Press, 1986. p. 216

25 Crawford DH, Epidemics and pandemics, doi:10.1093/actrade/9780198811718.003.0007, published in print March 22, 2018, published online March 22, 2018, accessed on March 26, 2020

26 Asymptomatic carriers are fueling the COVID-19 pandemic. Here's why you don't have to feel sick to spread the disease, available at www.discovermagazine.com/health/asymptomatic-carriers-are-fueling-the-covid-19-pandemic-heres-why-you-dont, accessed on March 27, 2020

27 Coronavirus disease 2019 (COVID-19) situation report – 3, WHO, available at www.who.int/docs/default-source/coronaviruse/situation-reports/20200227-sitrep-38-covid-19.pdf?sfvrsn=9f98940c_2, accessed on March 27, 2020

28 Coronavirus disease 2019 (COVID-19) situation report – 32, WHO, available at www.who.int/docs/default-source/coronaviruse/situation-reports/20200221-sitrep-32-covid-19.pdf?sfvrsn=4802d089_2, accessed on March 25, 2020

29 Coronavirus disease 2019 (COVID-19) situation report – 66, available at www.who.int/docs/default-source/coronaviruse/situation-reports/20200326-sitrep-66-covid-19.pdf?sfvrsn=81b94e61_2, accessed on March 27, 2020

30 Morawska L, Cao J, Airborne transmission of SARS-CoV-2: The world should face the reality, *Environment International*, June 2020; 139: 105730

31 Fineberg HV, *Rapid Expert Consultation on the Possibility of Bioaerosol Spread of SARS-CoV-2 for the COVID-19 Pandemic (April 1, 2020)*, Washington, DC: The National Academies Press, National Research Council, 2020

32 Coronavirus disease 2019 (COVID-19), CDC, available at www.cdc.gov/coronavirus/2019-ncov/php/water.html, accessed on March 27, 2020

33 Wang W, Xu Y, Gao R, et al., Detection of SARS-CoV-2 in different types of clinical specimens, *JAMA*, 2020, doi:10.1001/jama.2020.3786

34 Medema G, Heijnen L, Elsinga G, et al., Presence of SARS-coronavirus-2 in sewage, *Medrxiv*, 2020, doi:10.1101/2020.03.29.20045880

35 Heller L, Motab CR, Grecoc DB, COVID-19 faecal-oral transmission: Are we asking the right questions? *Science of the Total Environment*, August 10, 2020; 729: 138919

36 Coronavirus disease 2019 (COVID-19) situation report – 66

37 WHO, www.who.int/influenza/resources/documents/pandemic_phase_descriptions_and_actions.pdf, accessed on April 19, 2020

3

PANDEMICS

A trip down memory lane

COVID-19 is not the first pandemic to hit the world. In fact, humankind has faced the scourge of infectious diseases for thousands of years, and at one point of time, they were the leading cause of death across the world.

The scenario changed somewhat after the discovery of penicillin in 1928 by Sir Alexander Fleming.[1] Penicillin emerged as a new ammunition in the fight against infectious disease when physicians started using the drug actively in the 1940s. They put up a tough fight against various types of infectious diseases. Penicillin became one of the most sought-after medicines in the world during that period. However, despite decades of research and development, which has increased the capacity of physicians to treat and prevent infectious disease by leaps and bounds, infectious diseases remain a major cause of death and debility in the modern world.

With the development of highly advanced antibiotics, some physicians thought that infectious diseases would soon be eliminated and be of historic interest only. Newer medicines were developed which were found to be effective against viruses, fungi and other types of pathogens. However, these microbes were able to fight against such medicines. Antibiotic resistance developed fast and matched pace with the development of new antibiotics. Many diseases which experts once believed could be eliminated re-emerged with tremendous ferocity. Tuberculosis is a prime example.[2]

In recent times, the frequent interaction between humans and animals has added a new dimension to the development of infectious diseases. Disease-causing organisms which were previously confined to animals have now started causing infections in humans. Changes in environmental factors and proximity of human beings and animals are thought to be responsible for this alarming development. A brief review of the deadly pandemic that caused widespread death and debility in the past shows the destructive potential of these new diseases which left the medical science clueless about their cure or even prevention. Such episodes also prove

DOI: 10.4324/9781003345091-3

that we will most likely continue to see emergence of similar deadly diseases in the future as well and medical fraternity will find it difficult to deal with such health disasters whenever and wherever it appears, and no amount of preparedness would be enough to control such outbreaks. Nevertheless, we should try to learn from these incidents and do our best to at least develop emergency strategies which can help us minimize the impact of these infectious diseases.

Spanish flu

The Spanish flu started in 1918 and according to *Bulletin of the History of Medicine*, the pandemic resulted in about 50 million deaths all over the world.[3] Influenza is commonly known as the flu. It is an acute respiratory illness caused by influenza viruses A and B. New strains of influenza for which people have no immunity appear periodically, at irregular intervals, causing worldwide pandemics affecting vast numbers of people within short timespans. There have been 31 documented influenza pandemics since the first well-described pandemic of 1580, including three pandemics during the 20th century (1918, 1957 and 1969).[4] According to the WHO, the pandemic of 1918–1919, called the Spanish flu, was particularly virulent and killed an estimated 40 million people worldwide.[5]

Despite its name, the Spanish flu did not originate in Spain. It most likely originated in the United States. It was so called because Spanish newspapers were the first to report the outbreak. Spain was a neutral nation during World War I, and its journalists did not face strict censorship, unlike those working in countries involved in the war. The microbe circled the entire globe in four months, claiming millions of lives. The United States lost 675,000 people to the Spanish flu in 1918, more casualties than World War I, World War II, the Korean War and the Vietnam War combined.[6]

American forts experienced a massive expansion as the country mobilized for war. In Fort Riley, Kansas, a new training facility, Camp Funston, which was built to house some of the 50,000 men who would later be inducted into the army, was devastated by the Spanish flu. It was here in early March that a feverish soldier reported to the infirmary. Within a few hours, more than a hundred other soldiers had come down with a similar condition, and more would fall ill over the following weeks.[7] Once the disease subsided in the United States, it reached the battlefield of Europe. Americans fell ill with 'three-day fever' or 'purple death'. The French caught 'purulent bronchitis'. The Italians had 'sand fly fever'. German hospitals filled with victims of Blitzkatarrh or 'Flanders fever'.[8] Whatever they called it, the features were similar. It all started with fever and sore throat and chills and rapidly involved the lungs. Fluid poured into the lungs due to severe inflammation, and patients died due to respiratory failure. Lungs were full of fluid, and patients drowned in their own lung secretions.

The flu adversely affected the American population, with high fatalities. In the United States, stories abounded of people waking up sick and dying on their way to work, of overcrowded hospitals and of mass graves.[9]

Gradually, the flu reached the shores of India, China and even New Zealand. The pandemic left almost no part of the world untouched. It is estimated that in the United Kingdom, 228,000 people died; in Japan, some 400,000 people died.[10] The South Pacific island of Western Samoa (modern-day Samoa) lost about one-fifth of its population. Researchers estimate that in India alone, fatalities totalled between 12 million and 17 million. Although exact data are absent, global mortality figures are estimated to have been between 10% and 20% of those who were infected.[11]

Unfortunately, the greatest massacre of the 20th century was buried in the sands of time. Hardly anyone remembers it today as a pandemic; it is perhaps only remembered by a few people as private tragedies. However, the search for this deadly serial killer continued. Researchers extracted lung tissue from the body of a woman buried in Alaska deep under the layers of ice. She had died of the Spanish flu.[12] They succeeded in 2005, isolated the influenza virus, grew it in cell culture and studied its genome. The virus was found to be of avian origin, which jumped the species barrier and caused the deadly Spanish flu. They studied this virus in mice and found that its multiplication rate was much higher than that of other flu viruses and that it attacked the lungs specifically, where its concentration was 39,000 times greater than that of other flu viruses.[13] Dr Tumpey and colleagues wrote that 'the constellation of all eight genes together make an exceptionally virulent virus'.[14]

SARS

Severe acute respiratory syndrome (SARS) appeared in November 2002 in the Guangdong province of southern China. It was a viral respiratory illness caused by a new strain of coronavirus. The number of cases increased, and it turned into an epidemic in 2002–2003, involving four continents and 87 countries and causing about 744 deaths.[15] However, according to the WHO, SARS affected 26 countries and resulted in more than 8000 cases in 2003.[16] The difference between the data provided by WHO and other authorities is most likely due to different methods of case diagnosis, data collection and reporting.

The SARS coronavirus (SARS-CoV) was identified in 2003. SARS-CoV is thought to be an animal virus from an as-yet-uncertain animal reservoir, perhaps bats, that spread to other animals (civet cats).[17] Similar to COVID-19, SARS is thought to be transmitted most readily by respiratory droplets (droplet spread) produced when an infected person coughs or sneezes. The virus also can spread when a person touches a surface or object contaminated with infectious droplets and then touches their mouth, nose or eye(s). In addition, it is possible that the SARS virus might spread more broadly through the air (airborne spread) or by other ways that are not yet known.[18] Most cases of human-to-human transmission occurred in hospitals and other healthcare settings, due to a lack of adequate infection control precautions.

The common symptoms in SARS were fever (usually higher than 38 degrees centigrade), malaise, myalgia, headache, diarrhoea and shivering. Fever was the most frequently reported symptom, but sometimes it was absent, especially in elderly and immunosuppressed patients. Cough, breathlessness and diarrhoea were present in the first and/or second week of illness. Most patients developed pneumonia. Severe cases often evolved rapidly, requiring critical care in an ICU.[19]

Many features of COVID-19 are similar to that of SARS. COVID-19 and SARS have a median incubation time of about five days. Risk factors for severe disease outcomes, including death, ICU admission and mechanical ventilation, are old age and comorbidities. The progression for patients with severe disease follows a similar pattern in both the infections. Lung abnormalities on chest CT scans show the greatest severity approximately ten days after the initial onset of symptoms.[20]

In terms of severity, however, SARS is considered to be more deadly than COVID-19. The *British Medical Journal* mentioned that the WHO's director-general, Tedros Adhanom Ghebreyesus, said during a press briefing that more than 80% of patients with COVID-19 have a 'mild disease and will recover' and that it is fatal in 2% of reported cases. In comparison, the 2003 outbreak of SARS had a case fatality rate of around 10% (8098 cases and 774 deaths).[21]

However, COVID-19 seems to be much more infectious than SARS. Asymptomatic people are thought to be responsible to a great extent in spreading COVID-19. A recent article published in the *Journal of Emerging Infectious Diseases*[22] has estimated R_0 in COVID-19 to be as high as 5.7, much greater than that of SARS, which is supposed to vary between 2 to 4. The highly infectious nature of COVID-19 has resulted in its global spread within a few months of the outbreak.

Despite being a more fatal infectious disease, the SARS outbreak ended with implementation of appropriate and effective infection control practices, and no fresh cases have been reported since 2004. To date, there is no effective vaccine against it. Treatment is mainly supportive care. There is no reason why we cannot expect a similar course for COVID-19 as well.

However, there are a few lessons which should be learnt from the SARS epidemic, which was the first pandemic of the 21st century. It gave the world a warning which should have alerted public health professionals, lawmakers and economists to the possibility of a similar outbreak in the future. Retrospectively, it can be viewed as a prelude to the COVID-19 pandemic with more or less similar clinical features and socioeconomic disruptions. SARS had clearly demonstrated our vulnerability to infectious disease and forewarned the world to prepare for worse.

Swine flu

Swine flu started in 2009, spread almost all over the world and caused about 151,700–575,500 deaths.[23] The infection was detected first in the United States and spread quickly across the United States and the world. The pathogen responsible for it was a novel influenza A(H1N1) virus which had a new combination of influenza genes unidentified in humans or animals. This virus was named influenza A(H1N1)pdm09 virus.[24]

Because it was a new virus, few people had any immunity against it. However, about one-third of people older than 60 years had antibodies against this virus, likely from exposure to an older H1N1 virus earlier in their lives.

The disease affected mainly children, other young people and middle-aged adults. The clinical manifestations were just like other typical seasonal influenza. There was fever, cough, runny nose, sore throat, body aches, headaches, fatigue and more. Pneumonia with respiratory failure was the critical feature which was observed in severely ill patients. Respiratory complications were the main cause of death from swine flu worldwide.

By May 7, 2009, the disease was spreading rapidly, with the number of confirmed cases rising to 2099. On June 11, 2009, the WHO declared the 2009 H1N1 influenza a pandemic.[25] From April 12, 2009, to April 10, 2010, the CDC estimated that there were 60.8 million cases (range: 43.3–89.3 million), 274,304 hospitalizations (range: 195,086–402,719) and 12,469 deaths (range: 8868–18,306) in the United States due to the (H1N1)pdm09 virus. Globally, 80% of (H1N1)pdm09 virus–related deaths were estimated to have occurred in people younger than 65, whereas for seasonal influenza, maximum death is observed in people older than 65.

On August 10, 2010, the WHO declared an end to the global 2009 H1N1 influenza pandemic. However, the (H1N1)pdm09 virus continues to circulate as a seasonal flu virus and causes illness, hospitalization and deaths worldwide every year.

MERS

Another coronavirus-induced epidemic hit the world in 2012. Middle East respiratory syndrome (MERS) was an illness caused by a coronavirus called Middle East respiratory syndrome coronavirus (MERS-CoV).[26] The MERS epidemic which started in the Arabian Peninsula in 2012 spread over 22 countries and caused 659 deaths.[27] According to the WHO, since September 2012, 2494 laboratory-confirmed cases of infection with MERS-CoV were reported, and 858 MERS-CoV-associated deaths occurred across 27 countries.[28]

The first case of MERS was reported in Saudi Arabia in September 2012. However, through retrospective investigations, it was identified that the first-known cases of MERS occurred in Jordan in April 2012. According to the CDC so far, all cases of MERS have been linked through travel to, or residence in, countries in and near the Arabian Peninsula. The largest known outbreak of MERS outside the Arabian Peninsula occurred in the Republic of Korea in 2015. The outbreak was associated with a traveller returning from the Arabian Peninsula.[29]

Humans are infected with MERS-CoV from direct or indirect contact with dromedary camels. The clinical features of MERS-CoV infection has been reported as ranging from asymptomatic infection to acute upper respiratory illness and rapidly progressive pneumonia, respiratory failure, septic shock and multi-organ failure resulting in death. The common features are fever, dry cough, breathlessness, sore throat, fatigue, headache and more. The median incubation period is approximately five days (range: two to 14 days). The majority of the cases are adult, although children and adults of all ages have been infected (range: 0 to 109 years). Most hospitalized MERS-CoV patients had chronic comorbidities, just like the present COVID-19 cases. In critically ill patients, the median time from onset to ICU admission is approximately five days, and median time from onset to death is approximately 12 days. According to the CDC, the case fatality proportion is approximately 35%, which is one of the highest among recent epidemics.[30]

MERS-CoV has spread from ill people to others through close contact, such as caring for or living with an infected person. There is no specific treatment for MERS. Treatment is supportive care and the implementation of control and prevention measures. Thus far, no sustained human-to-human transmission has occurred anywhere in the world, but limited unsustained human-to-human transmission in healthcare facilities remains a prominent feature of this virus.

Comparison of SARS, MERS and COVID-19

The three coronavirus pandemics in this century have caused immense death and debility (Figure 3.1). The three viruses are more or less similar with more or less similar clinical manifestations. However, the infectivity and virulence differ. It is therefore prudent to compare the three pandemics to get an overall picture of coronavirus pandemics (Table 3.1).

Ebola Virus Disease

According to the CDC, Ebola virus disease (EVD) was discovered in 1976 when two outbreaks of fatal haemorrhagic (bleeding) fever occurred in different parts

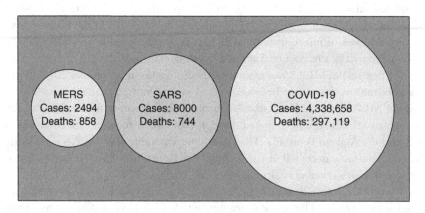

FIGURE 3.1   Graphic Representation of Total Number of Cases and Deaths of SARS, MERS and COVID-19 by May 15, 2020

Source: WHO.

TABLE 3.1 Comparison of SARS, MERS and COVID-19

Features	SARS	MERS	COVID-19
Infectious Agent	Beta-coronavirus	Beta-coronavirus	Beta-coronavirus
Name of Virus	SARS-CoV	MERS-CoV	SARS-CoV-2
Incubation Period	2 to 7 days	2 to 14 days	2 to 14 days
Main Symptoms	Fever, cough, shortness of breath	Fever, cough, shortness of breath	Fever, cough, shortness of breath
Transmission	Human to human	Human to human	Human to human
Mode of Transmission	Direct contact and droplet spread	Direct contact and droplet spread	Direct contact and droplet spread
Infectivity	Highly infectious	Highly infectious	Highly infectious
Cause of Death	Pneumonia, ARDS, MODS	Pneumonia, ARDS, MODS	Pneumonia, ARDS, MODS
Death Rate	About 10%	About 35%	About 2%
Countries Affected	26	27	All countries
Number of Cases	8000	2494	Ongoing pandemic: 1,914,916 cases detected (by April 15, 2020)
Number of Deaths	744	858	123,010 (by April 15, 2020)

Source: WHO.

of Central Africa.[31] The first outbreak occurred in the Democratic Republic of Congo (formerly Zaire) in a village near the Ebola River, which gave the virus its name. The second outbreak occurred in South Sudan, approximately 850 km away.

Scientists later discovered that the two outbreaks were caused by two genetically distinct viruses: *Zaire ebolavirus* and *Sudan ebolavirus*.[32] African fruit bats are likely involved in the spread of Ebola virus and may even be the source animal or reservoir.[33]

On March 23, 2014, the WHO reported cases of EVD in the forested rural region of southeastern Guinea. This 2014–2016 Ebola outbreak in West Africa spread to urban areas and across borders within weeks and became a global epidemic within months. On August 8, 2014, the WHO declared it as a public health emergency of international concern (PHEIC), which is designated only for events with a risk of potential international spread. Over the duration of the epidemic, EVD spread to seven more countries: Italy, Mali, Nigeria, Senegal, Spain, the United Kingdom and the United States. Later, secondary infection, mainly in a healthcare setting, occurred in Italy, Mali, Nigeria and the United States.[34] The West Africa Ebola virus disease epidemic started in 2013, spread over ten countries and led to 11,323 deaths.[35]

Patients with EVD generally had an abrupt onset of fever and symptoms, typically 8 to 12 days after exposure. Fever, fatigue, vomiting, diarrhoea and anorexia were the most common symptoms of the 2014 West African outbreak. Profound fluid losses from the gastrointestinal tract resulted in volume depletion, metabolic abnormalities (including hyponatraemia, hypokalaemia and hypocalcaemia), shock and organ failure. Bleeding was not universally present but could manifest later in the course as petechiae, ecchymosis/bruising or mucosal haemorrhage. Frank haemorrhage was less common. The CDC reported that patients with a fatal disease usually developed more-severe clinical signs early during infection and died typically between day six and day sixteen of complications, including multi-organ failure and septic shock. The case fatality proportion among patients with a known outcome in Guinea, Liberia and Sierra Leone was 70%.[36]

The US FDA (Food and Drug Administration) has not yet approved a specific treatment for Ebola virus disease. It approved the Ebola vaccine rVSV-ZEBOV (tradename Ervebo) on December 19, 2019.[37] The rVSV-ZEBOV vaccine is a single-dose vaccine regimen that has been found to be safe and protective against only the *Zaire ebolavirus* species of ebolavirus.[38] However, it may not provide 100% protection for everybody.

Zika

Zika virus is a mosquito-borne virus that was first identified in monkeys in Zika forest near Entebbe, Uganda, in 1947. In 1952, it was identified in humans, in Uganda and the United Republic of Tanzania.

The first outbreak of Zika virus disease was reported in the Island of Yap (Federated States of Micronesia, located northeast of Papua New Guinea) in 2007.[39] The patients developed fever, rash, conjunctivitis and joint pain. This was followed by a large outbreak of Zika virus infection in French Polynesia in 2013 and other countries and territories in the Pacific. A large outbreak of Zika virus disease was reported in Brazil, in March 2015. In October 2015, in Pernambuco, Brazil, an increase in the number of newborns with microcephaly was reported. Microcephaly is a birth defect in which a baby's head is smaller than expected when compared to babies of the same sex and age. Babies with microcephaly often have smaller brains that might not have developed properly. By the end of 2015, 4180 cases of suspected microcephaly had been reported.[40] Outbreaks soon appeared throughout the Americas, Africa and other regions of the world. To date, 86 countries and territories have reported evidence of mosquito-transmitted Zika infection.[41] On February 1, 2016, the WHO declared a PHEIC regarding clusters of microcephaly cases and neurological disorders in some areas affected by Zika virus.

Zika virus is mainly transmitted by the bite of an infected *Aedes aegypti*. This mosquito also transmits dengue, chikungunya and yellow fever. It usually bites during the day, peaking during early morning and late afternoon/evening. It is also transmitted from mother to foetus during pregnancy, through sexual contact, transfusions of blood and blood products, and organ transplantation.

The incubation period of Zika virus disease is about 3 to 14 days. The majority do not exhibit any symptoms, like COVID-19. Patients usually complain of fever, rash, muscle and joint pain, conjunctivitis, headache and malaise, which last for two to seven days.

The important complication of Zika virus disease is Guillain-Barré syndrome and microcephaly.[42] In Guillain-Barré syndrome, there is paralysis of the legs and arms due to damage of nerve fibres. Zika virus infection during pregnancy causes microcephaly (small head size) and other congenital abnormalities in the developing foetus and newborn. It also leads to foetal loss, stillbirth and preterm birth.

There is no vaccine or treatment for Zika virus disease. Protection from mosquito bites is the main method of prevention.

Retrospectively, all these pandemics exposed our vulnerability to biological pathogens and demonstrated the lack of efficacy that our healthcare systems have in dealing with pandemics. Careful planning, allocating funds for research and strengthening public health are necessary to limit the damage caused by pandemics, as we are still far away from preventing an outbreak of infection in the future which has the potential to turn into a pandemic within a short time period.

Pandemics spread rapidly across international borders. They start in a city or village and spread to other cities, and then the whole country is quickly affected. Global spread is associated with international travel. Pandemic control measures under the guidance of an international health organization like the WHO are implemented to limit the spread and damage caused by it. However, the implemented measures and their effectiveness vary from one country to another, and it is prudent to study the spread and global response to pandemics to limit their consequences.

Notes

1 Tan SY, Tatsumura Y, Alexander Fleming (1881–1955): Discoverer of penicillin, *Singapore Med J*, July 2015; 56(7): 366–367
2 Porter JD, McAdam KP, The re-emergence of tuberculosis, *Annu Rev Public Health*, 1994; 15: 303–323
3 Johnson NPAS, Mueller J, Updating the accounts: Global mortality of the 1918–1920 'Spanish' influenza pandemic, *Bulletin of the History of Medicine*, 2002; 76(1): 105–115
4 Report on global surveillance of epidemic-prone infectious diseases – Influenza, WHO, available at www.who.int/csr/resources/publications/influenza/CSR_ISR_2000_1/en/, accessed on March 27, 2020
5 Ibid.
6 Fujimura SF, Purple death: The great flu of 1918, available at www.paho.org/English/DD/PIN/Number18_article5.htm, accessed on March 27, 2020
7 Inside the swift, deadly history of the Spanish flu pandemic, *National Geographic*, available at www.nationalgeographic.com/history/magazine/2018/03-04/history-spanish-flu-pandemic/, accessed on March 27, 2020
8 Fujimura, Purple death: The great flu of 1918
9 Further suggested reading – *The Great Influenza*, book by John M. Barry
10 Inside the swift, deadly history of the Spanish flu pandemic, *National Geographic*
11 Ibid.
12 The deadliest flu: The complete story of the discovery and reconstruction of the 1918 pandemic virus, CDC, available at www.cdc.gov/flu/pandemic-resources/reconstruction-1918-virus.html, accessed on April 11, 2020

13 Ibid.
14 Tumpey TM, Basler CF, Aguilar PV, et al., Characterization of the reconstructed 1918 Spanish influenza pandemic virus, *Science*, October 2005; 310(5745): 77–80
15 Wang MD, Jolly AM, Changing virulence of the SARS virus: The epidemiological evidence, *Bulletin of the World Health Organization*, 2004; 82(7): 547–548
16 SARS (Severe Acute Respiratory Syndrome), WHO, available at www.who.int/ith/diseases/sars/en/, accessed on March 28, 2020
17 Ibid.
18 SARS basics fact sheet, CDC, available at www.cdc.gov/sars/about/fs-sars.html, accessed on March 28, 2020
19 Ibid.
20 Can we contain the COVID-19 outbreak with the same measures as for SARS? Prof Annelies Wilder Smith et al., available at www.thelancet.com/journals/laninf/article/PIIS1473-3099(20)30129-8/fulltext, accessed on April 20, 2020
21 www.bmj.com/content/368/bmj.m641, accessed on April 20, 2020
22 Sanche S, Lin YT, Xu C, et al., High contagiousness and rapid spread of severe acute respiratory syndrome coronavirus 2, *Emerg Infect Dis*, July 2020 [date cited], doi:10.3201/eid2607.200282, available at https://wwwnc.cdc.gov/eid/article/26/7/20-0282_article, accessed on April 20, 2020
23 2009 H1N1 pandemic (H1N1pdm09 virus), CDC, available at www.cdc.gov/flu/pandemic-resources/2009-h1n1-pandemic.html, accessed on May 16, 2020; Dawood FS, Iuliano AD, Reed C, et al., Estimated global mortality associated with the first 12 months of 2009 pandemic influenza A H1N1 virus circulation: A modelling study, *The Lancet Infectious Diseases*, 2012; 12(9): 687–695
24 2009 H1N1 pandemic (H1N1pdm09 virus), CDC
25 World now at the start of 2009 influenza pandemic, WHO, available at www.who.int/mediacentre/news/statements/2009/h1n1_pandemic_phase6_20090611/en/, accessed on May 16, 2020
26 Middle East Respiratory Syndrome (MERS), CDC, available at www.cdc.gov/coronavirus/mers/about/index.html, accessed on March 28, 2020
27 Arabi YM, Balkhy HH, Hayden FG, et al., Middle East respiratory syndrome, *New England Journal of Medicine*, 2017; 376(6): 584–594
28 Middle East Respiratory Syndrome Coronavirus (MERS-CoV), WHO, available at www.who.int/emergencies/mers-cov/en/, accessed on March 28, 2020
29 Middle East Respiratory Syndrome (MERS), CDC
30 MERS clinical features, CDC, available at www.cdc.gov/coronavirus/mers/clinical-features.html, accessed on May 16, 2020
31 History of Ebola virus disease, CDC, available at www.cdc.gov/vhf/ebola/history/summaries.html, accessed on May 16, 2020
32 Ibid.
33 Ibid.
34 2014–2016 Ebola outbreak in West Africa, CDC, available at www.cdc.gov/vhf/ebola/history/2014-2016-outbreak/index.html, accessed on March 28, 2020
35 2016 Ebola situation report. Weekly data report, WHO, April 15, 2020
36 Ebola virus disease (EVD) information for clinicians in U.S. healthcare settings, CDC, available at www.cdc.gov/vhf/ebola/clinicians/evd/clinicians.html, accessed on March 28, 2020
37 Prevention and vaccine, CDC, available at www.cdc.gov/vhf/ebola/prevention/index.html, accessed on May 16, 2020
38 Ibid.
39 Duffy MR, Chen T-H, Hancock WT, et al., Zika virus outbreak on Yap Island, Federated States of Micronesia, *N Engl J Med*, 2009; 360: 2536–2543
40 Teixeiria MG, Costa MCN, de Oliveira WK, et al., The epidemic of Zika virus – Related microcephaly in Brazil: Detection, control, etiology, and future scenarios, *Am J Public Health*, April 2016; 106(4): 601–605

41 Zika virus, WHO, available at www.who.int/news-room/fact-sheets/detail/zika-virus, accessed on May 7, 2020

42 Méndez N, Oviedo-Pastrana M, Mattar S, et al., Zika virus disease, microcephaly and Guillain-Barré syndrome in Colombia: Epidemiological situation during 21 months of the Zika virus outbreak, 2015–2017, *Arch Public Health*, November 2, 2017; 75: 65

4

ORIGIN SPREAD AND GLOBAL RESPONSE TO COVID-19

The COVID-19 pandemic has spread worldwide, has made millions of people sick and triggered an international response spearheaded by the WHO to halt its spread. From Wuhan, China, it spread like wildfire. The virus has now visited almost every country[1] in the world, bringing along death and debility (Figure 4.1). COVID-19 is presently ravaging the whole world without sparing anybody. Somehow or other,

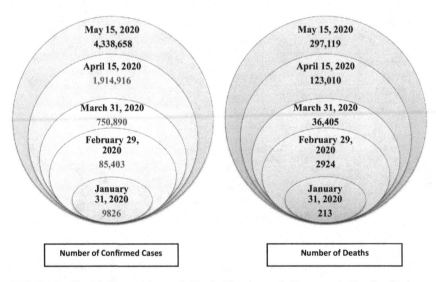

FIGURE 4.1 Rapid Progression of Total Number of Cases and Deaths Due to COVID-19

Source: WHO.

DOI: 10.4324/9781003345091-4

almost every person in this world has been affected. In a recent message, the WHO warned that the worst is yet to come.[2]

The best efforts by governments from every country have failed to halt its spread: cities were put under lockdown; people were advised to stay at home; international borders were closed; travel bans at local, national and international level were imposed; markets, schools, universities and shopping complexes were closed. Quarantine and self-isolation have been advised to stop the spread of COVID-19. The virus has triggered an unprecedented global crisis which led the WHO to provide technical guidance for government authorities, healthcare workers and other key stakeholders to respond to community spread.[3] It regularly updates the global situation on COVID-19[4] and at the same time coordinates with governments and various stakeholders to prepare for public health interventions to contain the pandemic.

The beginning

Wuhan, the capital of Hubei, China, is reported to be the place where the coronavirus first appeared and infected its first victim. But the Chinese authorities are not sure of its place of origin. This denial has resulted in confusion surrounding the origin of COVID-19. Many articles in various media have linked the Huanan Seafood Market at Wuhan with the origin of the virus, whereas some articles in media reported about several conspiracy theories regarding its origin.[5] Some considered it to be a human-made biological weapon created in a laboratory in China.[6] However, the conspiracy theory was refuted in an article published by the famous journal *Nature Medicine* on March 17, 2020, which said that it is not possible to artificially develop this kind of virus.[7]

First case of COVID-19

'I felt a bit tired, but not as tired as previous years. . . . Every winter, I always suffer from the flu. So I thought it was the flu'.[8] This is what W. G. said when she fell ill in December 2019. As reported by the *Wall Street Journal*, she is most likely the first case of this pandemic.[9] While selling shrimp at the Huanan Seafood Market, she felt sick and went to a local clinic, where she received an injection but did not recover. Ultimately, on December 16, she visited Wuhan Union Hospital, where she learnt that several other people from the Huanan Seafood Market had visited the hospital with similar symptoms. She was placed under quarantine and recovered later. By the time of her recovery, there were many others with cough and fever.

She was among the first 27 patients to test positive for COVID-19 and one of 24 cases with direct links to the market. However, a study published in the journal *Lancet* and reported by the BBC claims that the first case, identified on December 1, was an elderly man with Alzheimer's disease.[10] The *South China Morning Post* reported that the first case could be traced back to November 17.[11]

Whistle-blower Dr Li

Meanwhile, the novel coronavirus went on infecting people silently, and by the last week of December, there were too many victims to be ignored anymore by physicians in Wuhan. The patients came to the hospitals with cough and high fever, and X-rays revealed a new type of pneumonia. Many of these patients could not be linked directly to the seafood market. Experience of tackling SARS made the doctors wonder whether they were seeing another similar incidence. Whistle-blower Dr Li Wenliang,[12] an ophthalmologist, warned a group of other doctors about a possible outbreak of an illness that resembled SARS and urged his colleagues to use protective gear.

Wuhan Municipal Health Commission declared that investigations did not find any obvious human–to–human transmission and no medical staff infection.[13] Thus, the story ended on December 31, 2019. In first week of January 2020 the story took another turn, when Dr Li Wenliang was admonished by Wuhan Public Security Bureau, accusing the doctor of 'spreading rumours'.[14] On January 3, 2020, he was forced to sign a statement acknowledging his misdemeanour and promised not to indulge in any unlawful act.

Unfortunately, on January 10, Dr Li developed cough, and had fever the next day and was diagnosed with COVID-19 on January 30. Ultimately, on February 7, he breathed his last. This triggered massive online grief and rage against the Chinese government.[15] The tight-lipped ruling Communist Party was forced to issue a solemn apology to Dr Li's family. The two police officers were issued disciplinary punishments for their role in the original incident.[16]

Progression of cases between November 17, 2019, and March 31, 2020

By the time of the death of Dr Li, the virus had started spreading widely. According to the *South China Morning Post*, from November 17 to December 15, 2019, the total number of infections stood at 27 and by December 20, the total number of confirmed cases had reached 60.[17]

Dr Zhang Jixian, a doctor from Hubei Provincial Hospital of Integrated Chinese and Western Medicine, examined a senior couple and found them to have an unusual pneumonia and advised their son to see her.[18] She investigated him and diagnosed him with the same disease. Her experience in the management of SARS led her to think that this was caused by a new virus, and on December 27, Zhang Jixian informed China's health authorities that the mysterious disease was caused by a novel coronavirus. By December 31, 2019, the number of confirmed cases had risen to 266, and on January 1, 2020, they stood at 381. By February 1, 2020, the total number of confirmed cases in China totalled more than 11,000. By March 31, 2020, the total number of cases in China totalled 82,545, and the death toll was 3314.[19] On May 15, 2020, the total number of confirmed cases in China totalled 84,469, and the death toll had climbed to 4644.[20]

On January 1, 2020, the seafood market was closed, and workers in hazmat suits collected samples. However, the traders were kept in the dark and were asked to leave the market.[21] By January 20, the virus targeted the major cities of Beijing, Shanghai and Shenzhen.[22]

Discovery of SARS-CoV-2

The WHO China Joint Mission from February 16 to 24, 2020, declared their report and shed light on this mysterious disease.[23] The mission reported that the disease was caused by a new strain of coronavirus, which was later renamed SARS-CoV-2, and that it had affected 75,465 people all over China by February 20, 2020. Among 55,924 laboratory-confirmed cases, the median age was 51 years, with the majority of cases (77.8%) between ages 30 and 69. Among the reported cases, 51.1% were male, 77.0% were from Hubei and 21.6% were farmers or labourers by occupation.[24]

Strict measures were taken by the Chinese government to control the spread of the disease, including banning public gatherings, closing schools and malls and locking down Wuhan. However, all these measures failed to stop the spread of the coronavirus. By the second half of January 2020, the virus reached the neighbouring countries of South Korea and Japan and arrived in Europe and the United States. The WHO situation report dated January 21, 2020,[25] recorded the presence of COVID-19 cases in Thailand (two cases), Japan (one case) and South Korea (one case).

The story of South Korea

On January 20, 2020, nearly one month after COVID-19 was detected in China, the first case of COVID-19 was confirmed in South Korea.[26] In the fight against this deadly foe, South Korea organized one of the best programmes to contain the epidemic from spreading further. South Korea realized what had happened in Wuhan, and the country sprang into action even before the virus reached its shores. In a span of few days, the country prepared a test kit and started conducting tests when there were only a few cases.[27,28] The test kits were distributed in the regional health centres and were extensively used to identify even asymptomatic cases. The country used innovative measures for testing, like drive-through testing,[29] mobile applications for contact tracing and isolating the infected patients to stop further spread in the community.

The government also resorted to other precautionary steps to control the outbreak. By January 21, 2020, the government of South Korea scaled up the national alert level from Blue (Level 1) to Yellow (Level 2 out of a four-level national crisis management system). Similarly, quarantine and screening measures had been enhanced for travellers from Wuhan at the point of entries since January 3, 2020.[30] However, by February 20, there was a significant increase in the number of COVID-19 cases which was largely due to a gathering of over 500 people in

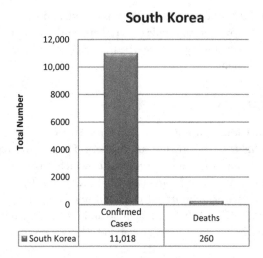

FIGURE 4.2a Total Number of Confirmed Cases and Deaths in South Korea Due to COVID-19 by May 15, 2020

Source: WHO.

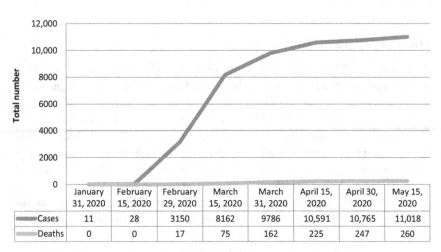

FIGURE 4.2b Steady Progression in the Total Number of Confirmed Cases and Deaths Due to COVID-19 in South Korea by May 15, 2020

Source: WHO.

the Shincheonji Church of Jesus at Daegu,[31] about 240 kilometres southeast of Seoul, where devotees from Wuhan were also present. A 61-year-old woman, who was later called Case 31, was present at that gathering and later tested positive for COVID-19.[32] She had attended two such gatherings and was believed to have infected numerous people, which resulted in a sharp rise in positive cases in South Korea. S. M. H., a 72-year-old woman, was among those who attended the church service. She developed severe cough with phlegm a few days later and could not sleep at all. Due to the absence of hospital beds in Daegu, she was taken to Seong-nam about 220 kilometres away and admitted to the hospital. After reaching the hospital, she said, 'I was relieved when I entered the hospital room – because at least there, I knew I wouldn't have died alone'.[33]

With rising positive cases of COVID-19 in the country, South Korea declared the highest possible alert on February 23, 2020. By February 29, the number of cases had reached 3150. After a week, on March 8, 2020, South Korea announced that the outbreak associated with Shincheonji Church totalled 4482 infections, accounting for 62.8% of the total confirmed cases.[34] On March 13, 2020, the number of recoveries was more than the number of newly tested positive cases.[35]

Another similar incident happened between March 1 and March 8, 2020, when about 46 churchgoers were infected with the coronavirus after attending the River of Grace Community Church.[36] The infections were claimed to have been caused by spraying salt water into followers' mouths,[37] under the belief that this would protect them from the virus. Other church clusters have appeared in the cities of Suwon, Busan, Geochang and Bucheon.[38]

The number of cases continued to rise, and by March 31, 2020, 9786 people were infected, and the death toll climbed to 162. According to the WHO, by April 15, 2020, the total confirmed cases were 10,591, and the total number of deaths was 225.[39] On May 15, 2020, the total number of cases were 11,018, and the death toll had climbed to 260 (Figure 4.2a and Figure 4.2b).[40]

South Korea has earned the praise of the global community for its rapid response, but it has earned its fair share of criticism as well. The extensive use of tracking applications and disclosure of the private information of infected people did not go over well with many.[41] However, South Korea cannot lower its guard. The pandemic is still evolving, and many people in South Korea are susceptible to coronavirus. There is no herd immunity yet. A vaccine is still far off. Hence, extensive testing, contact tracing, isolation and patient treatment should continue.

Invasion of Europe

Coronavirus arrived in Europe in late January and rapidly spread like wildfire from one country to another and spread all over Europe in next few weeks. The epicentre of the pandemic quickly shifted from Asia to Europe.[42] The reasons for the rapid spread were many. First, few actually thought the virus would land on their shores. It was considered to be a problem of East Asia. Second, the Western world and especially the medical community do not regularly deal with infectious diseases.

The virus did not give them any time to prepare. There was lack of test kits, medical masks, sanitizer, personal protective equipment (PPE) and so on.[43] Third, they did not opt for lockdown and the banning of air travel in the beginning. The curtailment of fundamental rights in rights-conscious citizens is equivalent to blasphemy. The end result has therefore been unprecedented death and debility in postwar Europe. The total number of confirmed cases and deaths as reported by the WHO on May 15, 2020, in various countries of the European region is depicted in Figure 4.3.[44]

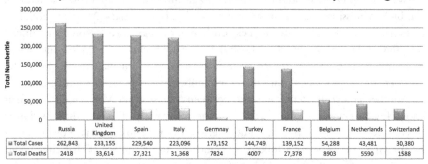

Total Number of Confirmed Cases and Deaths due to COVID-19 in Top Ten Worst-Affected Countries in the European Region

	Russia	United Kingdom	Spain	Italy	Germnay	Turkey	France	Belgium	Netherlands	Switzerland
Total Cases	262,843	233,155	229,540	223,096	173,152	144,749	139,152	54,288	43,481	30,380
Total Deaths	2418	33,614	27,321	31,368	7824	4007	27,378	8903	5590	1588

FIGURE 4.3   Total Number of Confirmed Cases and Deaths Due to COVID-19 in Top Ten Worst-Affected Countries of the European Region by May 15, 2020

Source: WHO.

France

In late January 2020, France became the first European country to confirm cases of coronavirus infection.[45] The case was confirmed at Bordeaux. He was a 48-year-old French citizen who arrived in France from China. France also recorded the first COVID-19 death in Europe. According to French Health Minister Agnès Buzyn, the victim was an 80-year-old Chinese tourist who died on February 14. He had arrived in France on January 16 and was admitted to a hospital. He became the first victim of the biological invader outside Asia.[46] A few days later, a 60-year-old school teacher became the first French national to die after developing COVID-19.

Meanwhile on February 8, 2020, Buzyn confirmed five new cases which originated from a group of people who were on a holiday in Les Contamines-Montjoie, Haute-Savoie.[47] They contracted the infection from a British national who had attended a conference in Singapore a few days earlier.

Like in South Korea, the annual assembly of the Christian Open Door Church between February 17 and 24, in Mulhouse, attended by about 2500 people, at least half of whom were believed to have contracted the virus, became one of Europe's largest regional clusters of infections, which then quickly spread across the country and eventually overseas. The French authorities recognized this cluster of cases late, and the virus had already spread by that time. In late February, multiple cases appeared in France, notably within three new clusters in Oise, Haute-Savoie and Morbihan.[48]

Surprisingly, the first round of municipal elections in France took place on March 15, 2020, against the backdrop of the raging epidemic.[49] The election with a voter turnout of only 45% was justified on the grounds that democratic values are the essence of French life. French nationals were supposed to, for hygienic reasons, bring their own ballpoint pen to fill out the ballot. The difficult balancing act between democratic values and public hygiene made it difficult for French nationals to appear before the ballot box.

Despair, fear, anxiety and loneliness gripped the people of France. French citizen C. B., 29, developed COVID-19 and was told, 'The tests are expensive and they keep them for the complex cases'.[50] He was married and had a 2-year-old son, and he was confined at his home in Paris. Fortunately, he improved gradually. Besides issues related to health, COVID-19 posed various other challenges as well. For a French homemaker and mother of three children, D. K., the worst part was the challenge that her children faced in continuing their schoolwork while confined at home.[51]

By March 1, 2020, the number of confirmed cases[52] eclipsed 100 in France, with two deaths, and within about next 20 days – i.e. by March 19, 2020 – the total number of cases jumped to 10,877, with 372 total fatalities. By the end of March, the virus claimed the lives of 3017 people while infecting 43,977 people. The number of infected people and deaths rose sharply thereafter, and by April 15, 2020, the total number of cases[53] came to 102,533, and the death toll had reached 15,708. On May 15, 2020, the total number of cases had increased to 139,152, with 27,378 deaths (Figure 4.4a and Figure 4.4b).[54] However, recently, the BBC reported that a patient treated in a hospital near Paris on December 27, 2019, for suspected pneumonia, actually had the novel coronavirus, as told by his treating doctor.[55] This means that the virus may have arrived in Europe almost a month earlier than previously thought. A swab taken at the time was recently tested and came back positive for COVID-19. The French health ministry told the BBC that the government was obtaining confirmation on the case and that it would consider further investigations if they proved necessary.

The virus went on, ultimately resulting in the deaths of thousands of French people. On March 16, President Macron announced a national lockdown for 15 days, starting on March 17 midday, and on March 27, Prime Minister Édouard Philippe announced that the lockdown would be extended until April 15.[56]

Lockdown has been strict in France. People have been asked to stay at home and venture out only for medical emergencies, to buy essential items and so on. Those ignoring the rules have been fined €135, and some repeat offenders have been jailed.

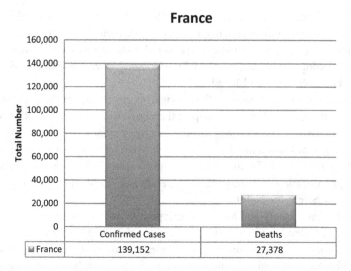

FIGURE 4.4a Total Number of Confirmed Cases and Deaths Due to COVID-19 in France by May 15, 2020

Source: WHO.

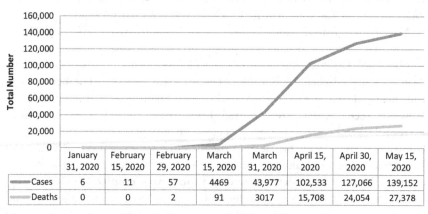

FIGURE 4.4b Rapid Progression in the Total Number of Confirmed Cases and Deaths Due to COVID-19 in France by May 15, 2020

Source: WHO.

Italy

Italy has been devastated by the deadly coronavirus. Prime Minister Giuseppe Conte of Italy announced the presence of the novel coronavirus in Italy after two Chinese tourists had been detected to be positive for COVID-19 in Rome on January 31.[57] They were hospitalized in isolation at the Spallanzani hospital in the capital. About a week later, an Italian national repatriated from Wuhan, China, was hospitalized and confirmed to have COVID-19. However, doubts have been raised regarding whether these were the first few cases in Italy. According to Reuters,[58] Italian researchers are looking into the higher than usual number of cases of severe pneumonia in Lombardy in the last quarter of 2019. According to Adriano Decarli, an epidemiologist and medical statistics professor at the University of Milan, hundreds of patients were hospitalized with pneumonia in the areas of Milan and Lodi between October and December 2019.[59] The virus had probably invaded Italy much earlier than last week of January 2020.

What happened at Lombardy?

Lombardy soon became the coronavirus hotspot. Around the middle of February, a 38-year-old man in Lombardy fell ill and was prescribed some treatment by a physician in Castiglione d'Adda.[60] He did not recover and went to Codogno Hospital, where he was further treated, but no one suspected COVID-19. Later, his wife said that he had met an Italian friend who had returned from China on January 21. Both the husband and the wife tested positive for coronavirus. They had met many people and led an active social life before diagnosis, and that enabled the virus to infect many people, including healthcare workers. Later, both of them were hospitalized. Gradually, the number of cases increased, and the first death was reported on February 22. The Lombardy cluster was greatly responsible for the spread of the virus in various parts of Italy. The highly contagious nature of the virus and lack of restriction on the movement of people allowed the infection to spread unchallenged. Ultimately, on March 8, Prime Minister Giuseppe Conte extended the quarantine lockdown to cover all the region of Lombardy and 14 other northern provinces.[61]

Game Zero

Experts believe that just two days before the detection of first COVID-19 case in Italy, a Champions League match between Atalanta and Valencia in February may have been the site from which the virus spread so rapidly in the country.[62] Atalanta hosted the Spanish team on February 19 in Milan. In the month of February, nobody was aware of the deadly virus circulating undetected. This game has been dubbed 'Game Zero' by the local media. Thousands of people were watching the game in the stands, sitting a few centimetres apart and then going to pubs and restaurants after the game. There were also about 2500 Valencia fans in the stands for the match, and 35% of Valencia's team tested positive for the novel coronavirus a few weeks later. A journalist who covered the match became the second person to test positive for COVID-19 in Bergamo. In addition, several Valencia fans became infected.[63] This football match was like a biological bomb that exploded later with devastating consequences.

Similarly, another cluster of cases appeared in Veneto, in northeast Italy. Gradually, the virus continued its deadly march across central and southern Italy, leaving behind an unprecedented number of casualties. In certain areas of northern Italy, the casualties were so high that there was hardly anybody to bury the dead. In Lombardy, Bergamo became the epicentre of COVID-19, in which there were hundreds of deaths. Families could not even give proper farewells to their dead relatives. Only a few minutes of funeral services were performed. The tragedy of COVID-19 is that while in hospital the patient could not meet their relatives because most of them were in quarantine. They died alone and were cremated with only a few near and dear ones around. The cemetery was so overwhelmed at Bergamo that military trucks were seen carrying the dead bodies to other places for burial. In certain villages and towns, there was hardly any family left untouched by the virus.

Grief, despair and hopelessness were widespread. No ray of light could be seen in the darkness. It was like living with the dead and dread. G. S., a COVID-19 survivor, left sub-intensive care and was brought, like many others, to a hotel in Bergamo, where nurses took care of her while she recovered. She said, 'When we'll be allowed out, I'll hug everybody and hold them very tight and let them understand how much I love them. My little great-grandson, I miss them all'.[64]

H. M., a 90-year-old man, and his wife tested positive for coronavirus. H. M. developed fever and later breathlessness. He was kept on a ventilator and recovered. 'I remember nothing from the night I almost passed to the other side', he said. His wife also recovered from the illness.[65]

By February 26, 2020,[66] there were 322 cases and 11 deaths, and by February 29, 2020, 888 cases were confirmed, with a total of 21 deaths. The virus went on claiming its victims at an alarming rate, and by March 11, 2020, the number of confirmed cases rose to 10,149, with 631 deaths. By March 31, 2020, the virus had infected 101,739 people and claimed the lives of 11,591 victims, which further rose to 162,488 positive cases and 21,069 deaths by April 15, 2020.[67] On May 15, the total number of cases increased to 223,096, and the death toll climbed to 31,368 (Figure 4.5a and Figure 4.5b).[68] No one is sure why Italy failed to contain the contagion. The casualties were much higher than many other places affected by COVID-19.

The Italian government sprang into action and suspended all flights to and from China and declared a state of emergency. In February, 11 municipalities in northern Italy were placed under quarantine; on March 8, 2020, Prime Minister Giuseppe Conte expanded the quarantine to all of Lombardy and 14 other northern provinces; and on the following day, he expanded it to all of Italy. On March 21, the government ordered the shutdown of all nonessential businesses and industries. Lockdown in Italy is strict, and police fine anyone who ventures out of their home without any valid reason. Prison riots have happened at some places during this period. Long queues are seen at the supermarkets.[69] However, these are probably the only means at present to control the spread of COVID-19.

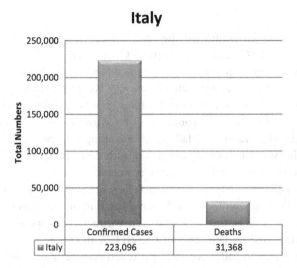

FIGURE 4.5a Total Number of Confirmed Cases and Deaths Due to COVID-19 in Italy by May 15, 2020

Source: WHO.

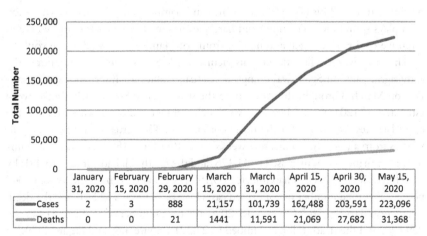

FIGURE 4.5b Rapid Progression in the Total Number of Confirmed Cases and Deaths Due to COVID-19 in Italy by May 15, 2020

Source: WHO.

Spain

On January 31, Spain confirmed the first case of COVID-19. As reported by CNN, a person in La Gomera, one of Spain's Canary Islands, was part of a group of five people who were in contact with a person infected with the novel coronavirus while they were in Germany.[70] On February 9, Reuters reported that a British man tested positive for coronavirus in Mallorca, thus becoming the second patient with COVID-19 in Spain. The patient was one of four members of a British family taken under observation in Mallorca after coming into contact with someone in France who was subsequently diagnosed with coronavirus infection. The test results for the man's wife and two daughters were negative.[71]

After 'Game Zero' in Milan, the Valencia fans were back in Spain without realizing the deadly baggage they carried home. Later, numerous fans tested positive. CNN reported that on February 27, a health department spokesperson in the Valencia region confirmed six new cases of COVID-19 in the city.[72] Three days later, a Portuguese man who had visited Valencia tested positive after returning home.

On February 24, an Italian national was kept in isolation at H10 Costa Adeje Palace Hotel, Tenerife, and the hotel was locked down. Besides him, four others also tested positive. Thus, the total number of active cases in the Canary Islands came to six: five in Tenerife and one in La Gomera. More than 700 tourists remained inside, waiting out for a 14-day isolation period.[73]

J. L., a 29-year-old resident of Seville, failed to escape from the clutches of the deadly virus. He had sheer panic wondering whether he could pass on the disease to his eight-week-old daughter. He was admitted to the ICU and ultimately fought off the virus to tell his tale. He said, 'Can you imagine, you're someone who is 29, who plays sports, doesn't smoke and hardly ever goes to the doctor, and you're in intensive care with oxygen and no one from your family is allowed to see you'.[74]

The virus was steadily attacking its victims while Spanish authorities ignored the early signs.[75] Life in Spain was going on as usual. Rallies for International Women's Day on March 8 brought thousands onto the streets across Spain, including a crowd estimated at 120,000 in Madrid. Two female cabinet ministers who attended the event later tested positive for the novel coronavirus. The virus went on claiming its victims from all over Spain in a few weeks' time. By March 18, 2020, the total number of confirmed cases increased to 11,178, and the death toll had climbed to 491; by March 31, 2020, the virus had infected 85,195 people and claimed the lives of 7340 people. The total number of confirmed cases by April 15 was 172,541, and the death toll increased to 18,056.[76] On May 15, 2020, the total number of confirmed cases was 229,540, and the death toll had climbed to 27,321 (Figure 4.6a and Figure 4.6b).[77]

Physicians in Spain had noticed pneumonia-like cases. They had observed what happened in Italy and informed the authorities. But everything fell on deaf ears.[78] As a result, all the warnings were overlooked, and no arrangements were made to procure face masks, gloves or PPE, and no efforts were made to upgrade the healthcare system to face the deluge of cases that would soon overwhelm the Spanish healthcare infrastructure. The Spanish government imposed a nationwide

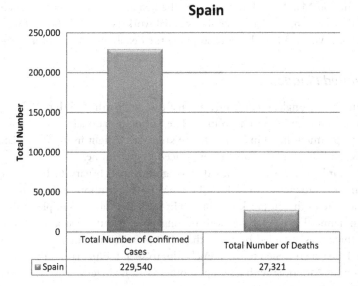

FIGURE 4.6a   Total Number of Confirmed Cases and Deaths Due to COVID-19 in Spain by May 15, 2020

Source: WHO.

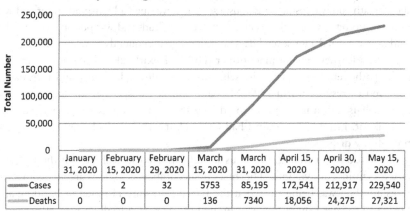

FIGURE 4.6b   Rapid Progression in the Total Number of Confirmed Cases and Deaths Due to COVID-19 in Spain by May 15, 2020

Source: WHO.

quarantine on March 14. On March 28, the Spanish government tightened up its national lockdown,[79] ordering all nonessential workers to stay at home for the next two weeks, with the faint hope of stopping the deadly march of coronavirus.

The United Kingdom

On January 31, England confirmed its first case in North Yorkshire.[80] Two people from the same family tested positive. The Chinese nationals were guests at the Staycity apartment hotel in York. On the same day, a flight from Wuhan landed in the UK, and all people on board were placed in quarantine.

The third patient was diagnosed at Brighton on February 6. He managed to pass on the coronavirus to at least 11 other people without ever setting foot in the epicentre of the outbreak in China. His story is a classic example of the spread of a pandemic due to globalization. In Singapore, he attended a business conference which was attended by more than 100 people, including one from Hubei, China. Later, he travelled to the French Alps to spend some days at the ski resort of Contamines-Montjoie with a group of other British citizens. Then he returned to the UK, with a cough and fever.[81]

Subsequently, on February 27, the first case was diagnosed in Northern Ireland, and the next day, another case was diagnosed in Wales – someone who had returned from Italy. On February 28, a resident of Surrey was diagnosed with a coronavirus infection who did not have any travel history. He was transferred to a specialist National Health Service (NHS) infection centre at Guy's and St Thomas' Hospital in central London.[82]A 51-year-old woman, L., in Chalfont St Peter in Buckinghamshire, had COVID-19. She said, 'I thought I was dying Wednesday last week, when I couldn't take a breath and was losing consciousness'. She developed a high-grade fever and was rushed to hospital, where she received intravenous fluids and was put on oxygen. Her condition stabilized after several days, and she was subsequently discharged.[83]

J., 59, a former general practitioner (GP) in Falkirk, also could not escape the deadly pathogen. She isolated herself from her vulnerable family and fortunately recovered to resume normal life. She said, 'I'm a week down the line from two weeks' self-isolation in a van on my driveway. . . . I have never experienced anything like it. I'm 99.5% certain I had COVID. It felt like I had a tight band going through my diaphragm'.[84]

In the month of March, the virus steadily affected several areas of the UK and started claiming its victims. The first to succumb was a 70-year-old woman in Berkshire. She was being treated at the Royal Berkshire Hospital in Reading and is believed to have caught the virus in the UK. The next to die was an 80-year-old man at Milton Keynes Hospital.[85] By March 21, 2020, the virus had attacked 5018 people and killed 233 people, and by March 31, 2020, 22,145 cases had been confirmed, with 1408 deaths. The WHO reported that the total number of confirmed cases had increased to 93,877 and that the death toll had also increased to 12,107 by April 15, 2020. Gradually, the number of infections and the death toll increased all over the UK.[86] On May 15, 2020, the total number of cases increased to 233,155, and the death toll had climbed to 33,614 (Figure 4.7a and Figure 4.7b).[87]

The United Kingdom

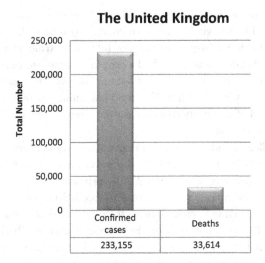

	Confirmed cases	Deaths
	233,155	33,614

FIGURE 4.7a   Total Number of Confirmed Cases and Deaths Due to COVID-19 in the United Kingdom by May 15, 2020

Source: WHO.

Rapid Progression of Cases and Deaths in the United Kingdom

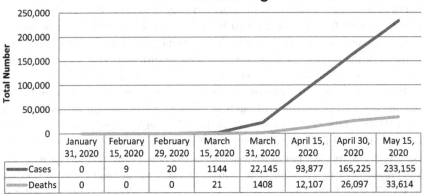

	January 31, 2020	February 15, 2020	February 29, 2020	March 15, 2020	March 31, 2020	April 15, 2020	April 30, 2020	May 15, 2020
Cases	0	9	20	1144	22,145	93,877	165,225	233,155
Deaths	0	0	0	21	1408	12,107	26,097	33,614

FIGURE 4.7b   Rapid Progression in the Total Number of Confirmed Cases and Deaths Due to COVID-19 in the United Kingdom by May 15, 2020

Source: WHO.

The response of the government was different from that of most of the countries that were affected by the viral menace. In contrast to others, the UK government chose not to shut down large gatherings or introduce stringent social-distancing measures. The officials instead described a plan to suppress the virus through gradual restrictions, rather than trying to stamp it out entirely. The strategy was to build herd immunity. They believed it to be a mild infection, one that would automatically raise the level of immunity to stamp out the virus if were to infect a significant proportion of population.[88]

Imperial College London urged that this strategy would overwhelm the healthcare facility, which led to a U-turn by the government, which introduced new social-distancing measures. On March 23, the prime minister ordered the closure of pubs, gyms and cinemas.[89] The Coronavirus Act 2020 received Royal Assent on March 25, having been fast-tracked through Parliament in just four sitting days.[90]

Meanwhile, Prime Minister Boris Johnson and Prince Charles, heir to the Royal Throne, fell prey to the deadly virus. Mr Johnson used his own case as a sort of teachable moment for the country, appealing to people to work from home and comply with the more drastic social-distancing measures that he had put in place.[91] He had to spend three nights in ICU at St Thomas' Hospital, and since his discharge, he has been recovering from COVID-19. The indiscriminate attack of the virus shook the entire nation, and it is still trying to recover from the shock.

Germany

On January 27, the first case of COVID-19 was confirmed to have been transmitted near Munich, Bavaria, in Germany. There was no history of travel to China, but there was contact with a Chinese guest visiting their company.[92] In Bavaria, the next cases were from German car parts supplier Webasto's headquarter in Stockdorf. The company reported that a Chinese employee from Shanghai had tested positive for the virus upon his return to China following his visit to the company headquarters.[93] Many cases later developed and were linked with Webasto.

By March 1, 2020, the number of confirmed cases was 129 in Germany without any deaths.[94] However, on March 8, an 83-year-old resident, diagnosed with COVID-19, of the St Nikolaus home of the elderly in Würzburg was brought into hospital and died four days later. He became the first casualty of COVID-19 in Bavaria. By March 27, ten more residents of the home died, and several residents and employees tested positive. The Würzburg old home became the virus hotspot of COVID-19 death. Many residents at old homes in all over Germany suffered the same fate.[95] The majority of these senior citizens also had serious comorbid conditions, including dementia.

There was another cluster of patients at Baden-Württemberg. On February 25, a 25-year-old man from the district of Göppingen, about 40 kilometres away from Stuttgart, had fallen ill with flu-like symptoms and had contacted the local health department. He was to be isolated and treated in a clinic. He had recently returned from Milan, Italy. The next day, a 24-year-old girl and her father in nearby

Tübingen were diagnosed with COVID-19. The girl happened to be the friend of the 25-year-old man and most likely contracted the disease from him.[96] The cases went on steadily increasing in Baden-Württemberg. The majority of these patients had contact with Italy, where they either travelled or came into contact with someone who had a history of travel to Italy.

North Rhine-Westphalia was attacked by the virus in its initial march through Germany. On February 25, a 47-year-old man tested positive in Erkelenz, Heinsberg. The next day, his wife tested positive as well. The couple attended carnival celebrations in the village of Langbroich and went on a short trip to the Netherlands. The man was seriously ill and was on ventilator support. The couple was being treated at the university hospital of Duesseldorf, the capital of North Rhine-Westphalia.[97] From these initial cases, the virus spread through several areas, and the case count started increasing. One of the infected people from the cluster in North Rhine-Westphalia was a doctor at Maria Hilf hospital in Mönchengladbach. After the case was identified, at least 35 people, including 15 of the doctor's patients, were medically examined.[98]

The virus's spread did not halt, and by March 16, at least 6600 cases of COVID-19 had been confirmed in Germany. Of them, 2400 were in North Rhine-Westphalia, more than 1000 in Bavaria and more than 900 in Baden-Wüerttemberg. At least 16 people in Germany were killed by the disease. By March 19, 2020, the total number of cases in Germany was 10,999, and the total number of fatalities was 20, and by March 31, 2020, the virus had infected 61,913 people and claimed the lives of 583 people. The total number of confirmed cases increased to 127,584, and the death toll had increased to 3254 by April 15, 2020.[99] On May 15, 2020, the total number of reported cases was 173,152, and the death toll had climbed to 7824 (Figure 4.8a and Figure 4.8b).[100]

Germany seems to be taking the epidemic in its stride with a high number of cases but a low number of deaths.[101] The mortality rate in Germany is far below its neighbouring European countries, and this has been put down to Germany's decision to implement widespread testing of people suspected of having the virus, as opposed to Italy's or the UK's decisions to test only symptomatic cases. The later arrival of the virus in Germany is supposed to be another factor. Besides this, the majority of patients were treated at top medical institutions. The slow increase in the number of cases is supposed to have helped in keeping the death toll low. Germany also has one of the best healthcare infrastructures in the world.[102]

In January, the German government did not think that the deadly virus was capable of causing harm to the German nation. By late January, the government was confident that it would be able to contain the spread of the virus.[103] Even in the middle of February, the government did not think of curtailing flights to and from China. By late February, the government started taking some steps, like closing schools and malls and regulating air and sea travel. On March 2, the Robert Koch Institute raised its threat level for Germany to moderate, and the European Centre for Disease Prevention and Control raised its threat level for Europe from moderate to high.[104]

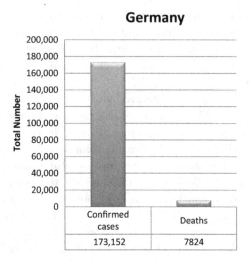

FIGURE 4.8a Total Number of Confirmed Cases and Deaths Due to COVID-19 in Germany by May 15, 2020

Source: WHO.

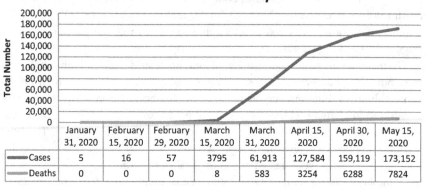

FIGURE 4.8b Rapid Progression in the Total Number of Confirmed Cases and Deaths Due to COVID-19 in Germany by May 15, 2020

Source: WHO.

In the middle of the crisis, on March 15, local elections in Bavaria took place. The government was still not serious about the pandemic. The number of cases and rising number of deaths, however, forced the government to take notice of the unfolding situation. On March 20, Bavaria was the first state to declare a curfew. The Bavarian government's move paved the way for other 15 states in Germany to impose their own stricter lockdowns or curfews. On March 22, the government and the federal states agreed for at least two weeks to forbid gatherings of more than two people and require a minimum distance of 1.5 metres (4 ft 11 in) between people in public.[105] Chancellor Angela Merkel's government has agreed with regional states to extend the nationwide lockdown aimed at slowing the spread of the coronavirus for another two weeks until April 19.[106]

Devastation in the US

The explosive and deadly global march did not halt in Europe. It crossed the Atlantic and reached the shores of the US. On January 19, 2020, a 35-year-old man came to a clinic in Snohomish County, Washington, complaining of cough and fever. He had a history of travel to Wuhan, China. A nasopharyngeal swab was collected, and the patient was discharged to home isolation with active monitoring by the local health department. On January 20, 2020, the US CDC confirmed that the swabs tested positive for COVID-19.[107]

A woman in her 60s became the second case of COVID-19 in the US when she returned to Chicago on January 13. She had returned to the US after taking care of her father in Wuhan, China.[108] A third case was confirmed a day later, in Orange County, California. This shows that within a few days, the virus had reached areas separated by hundreds of miles. On January 30, the first case of person-to-person transmission was confirmed. The husband of the second case did not travel to Wuhan but had frequent close contact with his wife after she returned from Wuhan. He was admitted eight days later and tested positive for SARS-CoV-2. Fortunately, both recovered from their respective illnesses.[109]

Almost every day, new cases were detected throughout the US as the virus spread rapidly through new territory at an alarming speed and ferocity. The virus killed its first victim, a man in his 50s, on February 29, admitted at Evergreen Health Medical Center in Kirkland, Washington, a suburb of Seattle.[110]

On March 1, 2020, in New York, Governor Andrew Cuomo announced the state's first reported case of COVID-19, a woman in her late 30s: a healthcare worker who apparently contracted the virus while travelling in Iran and was self-isolating at home, in New York City.[111] She experienced mild respiratory symptoms and went to Mount Sinai Hospital, where she tested positive for the novel coronavirus. New York City Health Commissioner Oxiris Barbot said the woman's neighbourhood was not at increased risk. 'We know that there's currently no indication that it's easy to transmit by casual contact. There's no need to do anything special in the community', Barbot said.[112] Unfortunately, she was wrong. The US was totally unprepared to face the virus. Mr Donald Trump, president of the US, throughout January, repeatedly played down the seriousness of the virus and focused on other issues.[113] According to CNN, Donald Trump downplayed the coronavirus threat 11 times in the early months.[114] The Central Intelligence Agency (CIA) issued several warnings, which he decided to ignore. Moreover, he ignored warnings conveyed in issues of the President's Daily Brief, a sensitive report that is produced before dawn each day and designed to call the president's attention to the most significant global developments and security threats.[115]

The virus spread throughout New York and went on attacking more and more people who had little defence against it. On March 14, it claimed its first victim in New York. An 82-year-old woman with a chronic lung disease developed COVID-19 and died after hospitalization.[116] A *Bloomberg News* editor was infected in the second week of March, and the odyssey of COVID-19 began for her. She had dry cough, high-grade fever, vomiting, sore throat, severe muscle pain and chills. A few days later, she became delusional and was rushed to the hospital. Her test turned out to be positive. A chest X-ray was conducted, which revealed the deadly pneumonia. However, she fought and overcame COVID-19. Her roommate had come weeks before she developed the symptoms. A few days afterwards, she too had similar symptoms.[117]

By the middle of March, mitigation measures expanded in New York, with a transition to online classes for universities and colleges and with the first semi-containment zone announced in New York. On March 20, Governor Andrew Cuomo issued a state-wide order that all nonessential workers must stay at home.[118] The same day, coronavirus cases in New York exceeded 7000.[119] By April 3, 2020, there have been at least 102,870 confirmed cases of the coronavirus discovered in New York, including more than 57,159 in New York City. At least 2935 people with COVID-19 have died in the state, which is about 38% of confirmed cases in the US.[120] How and why New York turned out to be the hotspot of coronavirus infection in New York is a matter of investigation that will enlighten the world to fight this deadly virus in coming days. However, the high density of its population, the highest rates of public transportation ridership and cold weather are the probable reasons.[121]

Although the US government failed to realize the gravity of the situation, it later moved swiftly to outpace the virus and save its citizens. President Trump on January 29, 2020, established the White House Coronavirus Task Force to coordinate and oversee efforts to 'monitor, prevent, contain, and mitigate the spread' of COVID-19 in the United States.[122] A series of travel restrictions was announced between late January and mid March. The entry of foreign nationals who had travelled through China, Iran, and 26 European countries that comprise Schengen area was denied. Later, the ban included travel to the UK and Ireland. American nationals had to undergo quarantine after health screening if they had travelled to those countries. In early March, the US CDC advised Americans against nonessential travel to countries affected by COVID-19.[123]

By March 21, governors in New York, California and other states had ordered most businesses to close and for people to stay inside.[124] Throughout March, several state, city and county governments imposed stay-at-home quarantines on their

United States of America

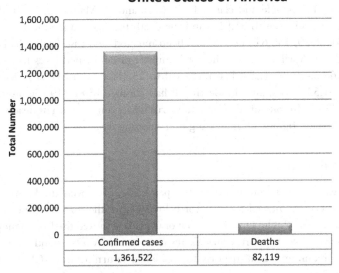

FIGURE 4.9a Total Number of Confirmed Cases and Deaths Due to COVID-19 in the United States by May 15, 2020

Source: WHO.

Rapid Progression of Cases and Deaths in United States of America

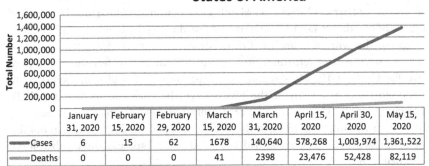

	January 31, 2020	February 15, 2020	February 29, 2020	March 15, 2020	March 31, 2020	April 15, 2020	April 30, 2020	May 15, 2020
Cases	6	15	62	1678	140,640	578,268	1,003,974	1,361,522
Deaths	0	0	0	41	2398	23,476	52,428	82,119

FIGURE 4.9b Rapid Progression in the Total Number of Confirmed Cases and Deaths Due to COVID-19 in the United States by May 15, 2020

Source: WHO.

populations to halt the spread of COVID-19. By March 19, 2020, the virus had infected 10,442 people and claimed 150 lives, and by March 25, 2020, the confirmed cases had risen to 51,914, and the death toll had climbed to 673.[125] As of March 31, 2020, 140,640 people had been infected, and case fatality had risen to 2398, and by April 15, 2020, the total number of confirmed cases had increased to 578,268, and the deaths had increased to 23,476.[126] On May 15, 2020, there were 1,361,522 cases, and the death toll had climbed to 82,119 (Figure 4.9a and Figure 4.9b).[127] In spite of all efforts, the virus is carrying on with its deadly activity, and the casualty figures are increasing every passing day.

Trouble in Canada

Canada was not left untouched by the pandemic. However, unlike many other countries, Canada braced for the storm long before the virus arrived. On January 15, 2020, the Public Health Agency of Canada activated the Emergency Operation Centre[128] to support Canada's response to COVID-19 and implemented screening requirements at airports for travellers returning from China on January 22, 2020.[129]

Even after taking necessary steps, Canada failed to stop the virus. It reared its ugly head by striking hard. The first case related to travel to Wuhan was confirmed on Canadian soil on January 25, 2020. A 56-year-old man in Toronto, Ontario, Canada, developed fever and dry cough, one day after returning from a three-month visit to Wuhan, China. He had fever, cough and bilateral pneumonia. He recovered and was discharged subsequently.[130] On February 20, 2020, Canada confirmed its first case related to travel outside mainland China.

A 77-year-old man with other coexisting health issues became the first victim of the virus. The man passed away at the Royal Victoria Regional Health Centre (RVH) in Barrie, Ontario, on March 11, according to Ontario's chief medical officer of health, Dr David Williams.[131] According to the BBC, Sophie Grégoire Trudeau, the wife of Canadian Prime Minister Justin Trudeau, tested positive for coronavirus after returning from a trip to London.[132] After she tested positive for COVID-19, Ms Grégoire Trudeau said, 'Although I'm experiencing uncomfortable symptoms of the virus, I will be back on my feet soon'. By March 31, 2020, the virus had already infected 6317 people and claimed the lives of 66 victims. According to the WHO, by April 15, 2020, the total number of confirmed cases in Canada had increased to 26,146, and the death toll had climbed to 823.[133]

On March 16, 2020, Canada advised travellers entering Canada to self-isolate for 14 days.[134] Canada took further steps to prevent the epidemic from spreading far, by banning the entry of foreign nationals from all countries, except the United States, on March 18, 2020. The Canada–US land border was closed to all nonessential travel. The best efforts of the government could not stop the virus. Thousands of people are now affected in Canada, with a few hundred dead. On May 15, 2020, there were 72,536 confirmed cases, with 5337 deaths (Figure 4.10a and Figure 4.10b).[135]

Canada

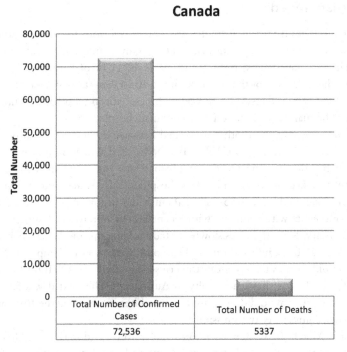

	Total Number of Confirmed Cases	Total Number of Deaths
	72,536	5337

FIGURE 4.10a Total Number of Confirmed Cases and Deaths Due to COVID-19 in Canada by May 15, 2020

Source: WHO.

Rapid Progression of Cases and Deaths in Canada

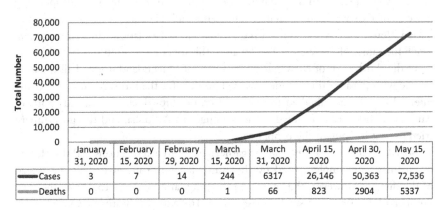

	January 31, 2020	February 15, 2020	February 29, 2020	March 15, 2020	March 31, 2020	April 15, 2020	April 30, 2020	May 15, 2020
Cases	3	7	14	244	6317	26,146	50,363	72,536
Deaths	0	0	0	1	66	823	2904	5337

FIGURE 4.10b Rapid Progression in the Total Number of Confirmed Cases and Deaths Due to COVID-19 in Canada by May 15, 2020

Source: WHO.

Australian tragedy

The far-flung continent of Australia was not spared from the novel coronavirus. Australia is no longer a remote place. The highly connected world allowed the lethal virus to travel across the sea and reach the coast of Australia. ABC News reported the detection of the first case in the Australian state of Victoria on January 25, 2020. The infected person was a Chinese citizen who had recently visited Wuhan. The man was admitted to a Melbourne hospital.[136]

On the same day, three other men aged between 30 and 60 tested positive for coronavirus, all in New South Wales. Two had visited Wuhan in recent weeks, and the other had direct contact with a confirmed case from Wuhan while in China. All three were kept in isolation in NSW hospitals.[137] By January's end, several cases had been diagnosed, and almost all of them had a history of travel to China or a history of contact with someone with recent history of travel to China.

In February 2020, the cases slowly started building up, and by late February, 24 Australians had been infected on the Diamond Princess cruise ship. On March 1, a 78-year-old man evacuated from the cruise ship died of COVID-19 at a hospital in Perth. He was the first casualty in Australia.[138] The next day, a 53-year-old healthcare worker, who did not have any travel history, became the first case to be infected by community transmission.[139]

In early March, B. W. flew back to Australia from London to attend her friend's wedding. But she never made it to the ceremony, because she contracted the virus. She spent the next 13 days in a Brisbane hospital. She said, 'It felt like a bit of a con, to be honest, to walk into hospital feeling fairly healthy, yet everyone's in hazmat suits'. Her condition deteriorated, and she felt very sick. However, she made a recovery and carried on with her life.[140]

By March 31, 2020, the virus had infected 4359 people and claimed the lives of 18 people. The killer virus did not care for reputation of a person before causing COVID-19. World-famous Hollywood actor Tom Hanks and his wife, Rita Wilson, were infected in mid March. They returned to Los Angeles after spending more than two weeks in quarantine in Australia.[141] Hanks wrote on Instagram, 'Good News: One week after testing positive, in self-isolation, the symptoms are much the same. No fever. . . . We are all in this together. Flatten the curve'.[142]

In the third week of March, the University of Queensland stopped all teaching for the week,[143] after three students tested positive for the virus. Meanwhile, many infected people disembarked from several cruise ships docked at various ports of Australia. The cases continued to rise throughout Australia, and the government tried its best to contain the pandemic.

Social distancing, banning public gatherings, border restrictions and the closure of cinemas were implemented by the Australian government to halt the spread of the disease. However, the cases are still rising, and by April 15, 2020, the total number of confirmed cases had risen to 6416, with 61 deaths.[144] On May 15, 2020, there was 6989 cases, and the death toll had climbed to 98 (Figure 4.11a and Figure 4.11b).[145]

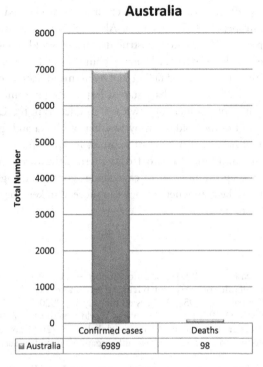

FIGURE 4.11a   Total Number of Confirmed Cases and Deaths Due to COVID-19 in Australia by May 15, 2020

Source: WHO.

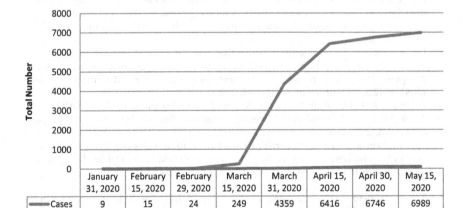

FIGURE 4.11b   Progression of Total Number of Cases and Deaths Due to COVID-19 in Australia by May 15, 2020

Source: WHO.

COVID-19 has caused widespread infection and has claimed several lives in several parts of Europe, the US and China. Although some countries could boast having the best possible healthcare infrastructure in the world, none could control the spread of the novel coronavirus. In many countries in Europe, the number of cases and the death toll have started falling, but by no means can it yet be concluded that the danger is over. Some cases have started reappearing in some parts of China. Still, there are significant numbers of new cases and deaths in the UK and the US. The infection started in the cold, clammy weather of China and spread rapidly to other countries that have similar types of weather. It now remains to be seen how it spreads in the hot and humid weather of the rest of the world. Many countries have been able to keep the number of cases under control during the initial few months. It remains to be seen whether they can succeed in keeping the virus under control in the days to come.

Notes

1 Coronavirus disease (COVID-19) situation report – 116, WHO, available at www.who.int/docs/default-source/coronaviruse/situation-reports/20200515-covid-19-sitrep-116.pdf?sfvrsn=8dd60956_2, accessed on May 16, 2020
2 The worst of COVID-19 is yet to come warns World Health Organization – video, *The Guardian*, available at www.theguardian.com/global/video/2020/apr/21/the-worst-of-covid-19-is-yet-to-come-warns-world-health-organization-video, accessed on May 15, 2020
3 Critical preparedness, readiness and response actions for COVID-19, WHO, available at www.who.int/emergencies/diseases/novel-coronavirus-2019/technical-guidance/critical-preparedness-readiness-and-response-actions-for-covid-19, accessed on May 15, 2020
4 www.who.int/ (the WHO website should be followed to learn about the latest developments regarding the coronavirus infection)
5 A bioweapon or effects of 5G? 7 conspiracy theories around coronavirus that will shock you, *The Economic Times*, available at https://economictimes.indiatimes.com/magazines/panache/is-covid-19-a-bioweapon-five-conspiracy-theories-around-coronavirus-that-will-shock-you/the-biggest-humanitarian-crisis/slideshow/74870568.cms, accessed on May 15, 2020
6 Ibid.; Pompeo says 'significant' evidence that new coronavirus emerged from Chinese lab, *Reuters*, available at https://in.reuters.com/article/health-coronavirus-usa-pompeo/pompeo-says-significant-evidence-that-new-coronavirus-emerged-from-chinese-lab-idINKBN22G04A, accessed on May 15, 2020
7 Andersen KG, Rambaut A, Lipkin WI, et al., The proximal origin of SARS-CoV, *Nature Medicine*, 2020, available at www.nature.com/articles/s41591-020-0820-9, accessed on April 8, 2020
8 Wuhan shrimp seller may be coronavirus 'patient zero' and says bug 'came from toilet', *MIRROR*, available at www.mirror.co.uk/news/world-news/wuhan-shrimp-seller-coronavirus-patient-21767997, accessed on April 2, 2020
9 Ibid.
10 Who is 'patient zero' in the coronavirus outbreak? *BBC*, available at www.bbc.com/future/article/20200221-coronavirus-the-harmful-hunt-for-covid-19s-patient-zero, accessed on May 17, 2020
11 *SCMP*, available at www.scmp.com/news/china/society/article/3074991/coronavirus-chinas-first-confirmed-covid-19-case-traced-back, accessed on April 2, 2020

12 Li Wenliang: Coronavirus kills Chinese whistleblower doctor, *BBC*, available at www.bbc.com/news/world-asia-china-51403795, accessed on May 15, 2020

13 Here's how China misled the world on coronavirus, *Pune Mirror*, available at https://punemirror.indiatimes.com/heres-how-china-misled-the-world-on-coronavirus/articleshow/74846484.cms, accessed on May 15, 2020

14 Li Wenliang: Coronavirus death of Wuhan doctor sparks anger, *BBC*, available at www.bbc.com/news/world-asia-china-51409801, accessed on April 2, 2020

15 Ibid.

16 Coronavirus: China apologises to family of doctor who died after warning about COVID-19, *Sky News*, available at https://news.sky.com/story/coronavirus-china-apologises-to-family-of-doctor-who-died-after-warning-about-covid-19-11960679, accessed on April 2, 2020

17 Coronavirus: China's first confirmed COVID-19 case traced back to November 17, by Josephine Ma.

18 Xinhua headlines: Chinese doctor recalls first encounter with mysterious virus, available at www.xinhuanet.com/english/2020-04/16/c_138982435.htm, accessed on May 15, 2020

19 Coronavirus disease 2019 (COVID-19) situation report – 71, WHO, available at www.who.int/docs/default-source/coronaviruse/situation-reports/20200331-sitrep-71-covid-19.pdf?sfvrsn=4360e92b_8, accessed on May 15, 2020

20 Coronavirus disease (COVID-19) situation report – 116, WHO

21 The Huanan Seafood Market was closed on January 1, according to *Business Insider*, available at www.businessinsider.in/slideshows/miscellaneous/photos-show-how-china-is-grappling-with-the-wuhan-coronavirus-outbreak-as-12-cities-are-quarantined-and-hospitals-run-out-of-space/slidelist/73598728.cms#slideid=73598751, accessed on May 15, 2020

22 New China virus: Cases triple as infection spreads to Beijing and Shanghai, *BBC*, available at www.bbc.com/news/world-asia-china-51171035, accessed on May 15, 2020

23 Report of the WHO-China joint mission on coronavirus disease 2019 (COVID-19), WHO, available at www.who.int/docs/default-source/coronaviruse/who-china-joint-mission-on-covid-19-final-report.pdf, accessed on April 2, 2020

24 Ibid.

25 Novel coronavirus (2019-nCoV) situation report – 1, WHO, available at www.who.int/docs/default-source/coronaviruse/situation-reports/20200121-sitrep-1-2019-ncov.pdf?sfvrsn=20a99c10_4, accessed on May 15, 2020

26 Novel coronavirus – Republic of Korea (ex-China), WHO, available at www.who.int/csr/don/21-january-2020-novel-coronavirus-republic-of-korea-ex-china/en/, accessed on May 15, 2020

27 Special Report: How Korea trounced U.S. in race to test people for coronavirus, *Reuters*, available at www.reuters.com/article/us-health-coronavirus-testing-specialrep/special-report-how-korea-trounced-u-s-in-race-to-test-people-for-coronavirus-idUSKBN-2153BW, accessed on May 15 2020

28 South Korea took rapid, intrusive measures against COVID 19 and they worked, *The Guardian*, available at www.theguardian.com/commentisfree/2020/mar/20/south-korea-rapid-intrusive-measures-covid-19, accessed on April 2, 2020

29 South Korea pioneers coronavirus drive-through testing station, *CNN*, available at https://edition.cnn.com/2020/03/02/asia/coronavirus-drive-through-south-korea-hnk-intl/index.html, accessed on May 15, 2020

30 Novel coronavirus – Republic of Korea (ex-China), WHO

31 South Korea's coronavirus cases rise to 6,767 with most cases traced to church, *Economic Times*, available at https://economictimes.indiatimes.com/news/international/world-news/south-koreas-coronavirus-cases-rise-to-6767-with-most-cases-traced-to-church/articleshow/74524331.cms?from=mdr, accessed on May 15, 2020

32 Coronavirus cases have dropped sharply in South Korea: What's the secret to its success? by Dennis Normile, available at www.sciencemag.org/news/2020/03/coronavirus-cases-have-dropped-sharply-south-korea-whats-secret-its-success#, accessed on April 21, 2020

33 Pain, solitude, fear: Survivors tell their stories of COVID 19, *The Japan Times*, available at www.japantimes.co.jp/news/2020/04/07/world/survivors-tell-their-stories-of-covid-19/#.XpBa4tR97Dc, accessed on April 10, 2020

34 [Corona Synthesis] One patient confirmed at Seoul Paik Hospital. . . 'Hide in Daegu and visit' [출처: 중앙일보], available at https://news.joins.com/article/23724705, accessed on May 15, 2020

35 South Korea reports 107 new coronavirus cases, total 8,086: KCDC, *Reuters*, available at www.reuters.com/article/us-health-coronavirus-southkorea/south-korea-reports-107-new-coronavirus-cases-total-8086-kcdc-idUSKBN21101Y, accessed on May 15, 2020

36 Coronavirus: Saltwater spray infects 46 church-goers in South Korea, *SCMP*, available at www.scmp.com/week-asia/health-environment/article/3075421/coronavirus-saltwater-spray-infects-46-church-goers, accessed on May 15, 2020

37 South Korea took rapid, intrusive measures against COVID 19 and they worked, *The Guardian*

38 Coronavirus cluster emerges at another South Korean church, as others press ahead with Sunday services, SCMP, available at www.scmp.com/week-asia/health-environ ment/article/3077497/coronavirus-cluster-emerges-another-south-korean, accessed on May 15, 2020

39 Coronavirus disease 2019 (COVID-19) situation report – 86, WHO, available at www.who.int/docs/default-source/coronaviruse/situation-reports/20200415-sitrep-86-covid-19.pdf?sfvrsn=c615ea20_6, accessed on May 15, 2020

40 Coronavirus disease (COVID-19) situation report – 116, WHO

41 South Korea is reporting intimate details of COVID-19 cases: Has it helped? *Nature*, available at www.nature.com/articles/d41586-020-00740-y, accessed on May 15, 2020

42 Coronavirus disease 2019 (COVID-19) situation report – 54, WHO, available at www.who.int/docs/default-source/coronaviruse/situation-reports/20200314-sitrep-54-covid-19.pdf?sfvrsn=dcd46351_8, accessed on April 9, 2020

43 EU fighting shortages and faulty medical supplies, *EUobserver*, available at https://euob server.com/coronavirus/147958, accessed on May 15, 2020

44 Coronavirus disease (COVID-19) situation report – 116, WHO

45 Coronavirus: First death confirmed in Europe, *BBC*, available at www.bbc.com/news/world-europe-51514837, accessed on April 3, 2020

46 Coronavirus: Infected Chinese tourist in France dies, in Europe's first death, *SCMP*, available at www.scmp.com/news/world/europe/article/3050802/coronavirus-infected-chinese-tourist-france-dies-europes-first, accessed on April 3, 2020

47 Coronavirus: British nine-year-old in hospital in France, *The Guardian*, available at www.theguardian.com/world/2020/feb/08/coronavirus-five-new-cases-in-france-are-british-nationals, accessed on May 15, 2020

48 2020 coronavirus pandemic in France, Wikipedia, available at https://en.wikipedia.org/wiki/2020_coronavirus_pandemic_in_France, accessed on April 3, 2020

49 France takes coronavirus precautions ahead of municipal elections, *France 24*, available at www.france24.com/en/20200314-france-introduces-coronavirus-precautions-ahead-of-municipal-elections, accessed on May 15, 2020

50 Pain, solitude, fear: Stories of surviving COVID-19, *Times of India*, available at https://timesofindia.indiatimes.com/world/europe/pain-solitude-fear-stories-of-surviving-covid-19/articleshow/75025118.cms, accessed on May 15, 2020

51 Ibid.

52 Coronavirus disease 2019 (COVID-19) situation report – 41, WHO, available at www.who.int/docs/default-source/coronaviruse/situation-reports/20200301-sitrep-41-covid-19.pdf?sfvrsn=6768306d_2, accessed on May 15, 2020

53 Coronavirus disease 2019 (COVID-19) situation report – 86, WHO
54 Coronavirus disease (COVID-19) situation report – 116, WHO
55 Coronavirus: France's first known case 'was in December', *BBC*, available at www.bbc. com/news/world-europe-52526554, accessed on May 17, 2020
56 French PM extends coronavirus lockdown until April 15, *France 24*, available at www. france24.com/en/20200327-french-pm-extends-coronavirus-lockdown-by-two-weeks-until-april-15, accessed on April 3, 2020
57 D'Arienzo M, Coniglio A, Assessment of the SARS-CoV-2 basic reproduction number, R_0, based on the early phase of COVID-19 outbreak in Italy, *Biosafety and Health*, available at www.sciencedirect.com/science/article/pii/S2590053620300410, accessed on May 15, 2020
58 Parodi E, Aloisi S, Italian scientists investigate possible earlier emergence of coronavirus, *Reuters*, available at www.reuters.com/article/us-health-coronavirus-italy-timing/italian-scientists-investigate-possible-earlier-emergence-of-coronavirus-idUSKBN21D2IG, accessed on April 3, 2020
59 Ibid.
60 Italy's coronavirus outbreak infects 51 people, kills two, *Reuters*, available at www.reuters. com/article/us-china-health-italy/second-coronavirus-patient-dies-in-italy-ansa-news-agency-idUSKCN20G0CQ, accessed on May 15, 2020
61 Italy has put the entire country under travel restrictions to curb coronavirus, available at www.vox.com/2020/3/9/21171373/italy-coronavirus-quarantine-lombardy-conte, accessed on May 15, 2020
62 Bengel C, Coronavirus: How a Champions League match contributed to Italy's COVID-19 outbreak, *CBS Sports*, available at www.cbssports.com/soccer/news/coronavirus-how-a-champions-league-match-contributed-to-italys-covid-19-outbreak/, accessed on April 3, 2020
63 Ibid.
64 COVID-19 survivors speak: 'You feel breathless and you understand it may be your last breath', *Euro News*, available at www.euronews.com/2020/04/01/covid-19-survivors-speak-you-feel-breathless-and-you-understand-it-may-be-your-last-breath, accessed on May 15, 2020
65 Ibid.
66 Coronavirus disease 2019 (COVID-19) situation report – 37, WHO, available at www.who.int/docs/default-source/coronaviruse/situation-reports/20200226-sitrep-37-covid-19.pdf?sfvrsn=2146841e_2, accessed on May 15, 2020
67 Coronavirus disease 2019 (COVID-19) situation report – 86, WHO
68 Coronavirus disease (COVID-19) situation report – 116, WHO
69 Leaked coronavirus plan to quarantine 16m sparks chaos in Italy, *The Guardian*, available at www.theguardian.com/world/2020/mar/08/leaked-coronavirus-plan-to-quarantine-16m-sparks-chaos-in-italy, accessed on May 15, 2020
70 First case of coronavirus confirmed in Spain, *CNN*, available at https://edition.cnn. com/asia/live-news/coronavirus-outbreak-02-01-20-intl-hnk/h_afcf3a4665521aab 11c721c8cc80dd03, accessed on April 3, 2020
71 Spanish authorities confirm Briton is country's second coronavirus case, *Reuters*, available at www.reuters.com/article/us-china-health-spain/spain-confirms-its-second-coronavirus-case-spanish-authorities-idUSKBN20309T, accessed on April 3, 2020
72 Lister T, Rebaza C, How Spain became a hotspot for coronavirus, *CNN*, available at https://edition.cnn.com/2020/03/28/europe/spain-coronavirus-hotspot-intl/index. html, accessed on April 3, 2020
73 New coronavirus case confirmed at Tenerife hotel on lockdown, *Reuters*, available at https://in.reuters.com/article/china-health-spain-tenerife/new-coronavirus-case-confirmed-at-tenerife-hotel-on-lockdown-idINKBN20O18K, accessed on April 3, 2020
74 www.ndtv.com/world-news/covid-19-update-worst-moment-of-my-life-says-spanish-coronavirus-survivor-29-2205239, accessed on April 10, 2020

75 Spain's coronavirus crisis accelerated as warnings went unheeded, available at www.nytimes.com/2020/04/07/world/europe/spain-coronavirus.html, accessed on May 15, 2020

76 Coronavirus disease 2019 (COVID-19) situation report – 86, WHO

77 Coronavirus disease (COVID-19) situation report – 116, WHO

78 Pinedo E, Faus J, As Spain battles virus, medics' unions hit out, *Reuters*, available at https://in.reuters.com/article/uk-health-coronavirus-spain-medics-insig/as-spain-battles-virus-medics-unions-hit-out-idINKBN21K0L5, accessed on April 3, 2020

79 Spain orders non-essential workers stay home for two weeks, *The Guardian*, available at www.theguardian.com/world/2020/mar/28/covid-19-may-be-peaking-in-parts-of-spain-says-official, accessed on May 15, 2020

80 First case of coronavirus confirmed in North Yorkshire: But no rise in York cases, *York Press*, available at www.yorkpress.co.uk/news/18301841.first-case-coronavirus-confirmed-north-yorkshire/, accessed on April 3, 2020

81 From Singapore to UK via the Alps: How one man spread coronavirus, *France 24*, available at www.france24.com/en/20200210-from-singapore-to-uk-via-the-alps-how-one-man-spread-coronavirus, accessed on April 3, 2020

82 Coronavirus: Latest patient was first to be infected in UK, *BBC*, available at www.bbc.com/news/uk-51683428, accessed on April 3, 2020

83 COVID-19 recoveries: 'It was the most terrifying experience of my life,' *The Guardian*, available at www.theguardian.com/society/2020/apr/01/covid-19-recoveries-it-was-the-most-terrifying-experience-of-my-life, accessed on April 10, 2020

84 Ibid.

85 Coronavirus: Man in 80s is second person to die of virus in UK, *BBC*, available at www.bbc.com/news/uk-51771815, accessed on April 3, 2020

86 Coronavirus disease 2019 (COVID-19) situation report – 86, WHO

87 Coronavirus disease (COVID-19) situation report – 116, WHO

88 The U.K. backed off on herd immunity: To beat COVID-19, we'll ultimately need it, *National Geographic*, available at www.nationalgeographic.com/science/2020/03/uk-backed-off-on-herd-immunity-to-beat-coronavirus-we-need-it/, accessed on April 3, 2020

89 Ibid.

90 www.instituteforgovernment.org.uk/explainers/coronavirus-act, accessed on April 3, 2020

91 Boris Johnson contracts coronavirus, rattling top ranks of U.K. Government, *New York Times*, available at www.nytimes.com/2020/03/27/world/europe/boris-johnson-coronavirus.html, accessed on April 3, 2020

92 European Centre for Disease Prevention and Control (ECDC). *ECDC Statement Following Reported Confirmed Case of 2019-nCoV in Germany*, Stockholm: ECDC, January 28, 2020, available at www.ecdc.europa.eu/en/news-events/ecdc-statement-following-reported-confirmed-case-2019-ncov-germany, accessed on January 30, 2020

93 German auto supplier Webasto says two employees infected with coronavirus, available at https://auto.economictimes.indiatimes.com/news/auto-components/german-auto-supplier-webasto-says-two-employees-infected-with-coronavirus/73721439, accessed on April 3, 2020

94 Coronavirus disease 2019 (COVID-19) situation report – 41, WHO

95 www.kxan36news.com/ticking-time-bombs-in-the-virus-crisis-retirement-homes-to-corona-hotspots, accessed on April 3, 2020

96 Update: Five new coronavirus cases confirmed in western Germany, available at www.thelocal.de/20200226/two-new-coronavirus-cases-confirmed-in-germany, accessed on April 4, 2020

97 Carnival-going German couple contract coronavirus, authorities fear it has spread, *Reuters*, available at https://in.reuters.com/article/china-health-germany/carnival-going-german-couple-contract-coronavirus-authorities-fear-it-has-spread-idINKCN20K2E5, accessed on April 4, 2020

98 Outbreak of novel coronavirus (COVID-19) – Germany, available at https://global-monitoring.com/gm/page/events/epidemic-0001937.eP207vO5rIwe.html?lang=en, accessed on April 4, 2020

99 Coronavirus disease 2019 (COVID-19) situation report – 86, WHO

100 Coronavirus disease (COVID-19) situation report – 116, WHO

101 Ellyatt H, Germany has a low coronavirus mortality rate: Here's why, *CNBC*, available at www.cnbc.com/2020/04/03/germany-has-a-low-coronavirus-mortality-rate-heres-why.html, accessed on May 17, 2020

102 Ibid.

103 Germany pulls out 'bazooka' against the coronavirus – But is it doing enough? available at www.brookings.edu/blog/order-from-chaos/2020/03/17/germany-pulls-out-the-bazooka-against-the-coronavirus-but-is-it-doing-enough/, accessed on May 15, 2020

104 EU raises risk level of coronavirus infection as global deaths pass 3,000, *The Guardian*, available at www.theguardian.com/world/2020/mar/02/eu-raises-risk-coronavirus-infection-from-moderate-high, accessed on April 4, 2020

105 2020 coronavirus pandemic in Germany, Wikipedia, available at https://en.wikipedia.org/wiki/2020_coronavirus_pandemic_in_Germany, accessed on April 3, 2020

106 Donahue P, German government extends nationwide lockdown until April 19, available at www.bloomberg.com/news/articles/2020-04-01/german-government-extends-nationwide-lockdown-until-april-19, accessed on April 4, 2020

107 Holshue ML, DeBolt C, Lindquist S, et al., First case of 2019 novel coronavirus in the United States, *NEJM*, available at www.nejm.org/doi/full/10.1056/NEJMoa2001191, accessed on April 4, 2020

108 Chicago woman who travelled to China diagnosed with coronavirus, health officials say, *The Chicago Tribune*, available at www.chicagotribune.com/news/breaking/ct-coronavirus-china-epidemic-illinois-case-20200124-yx2xd3yeovar3o25ei6bfvvbze-story.html, accessed on April 4, 2020

109 Ghinai I, McPherson TD, Hunter JC, et al., First known person-to-person transmission of severe acute respiratory syndrome coronavirus 2 (SARS-CoV-2) in the USA, *Lancet*, March 12, 2020, available at www.thelancet.com/action/showPdf?pii=S0140-6736%2820%2930607-3, accessed on April 4, 2020

110 First US death from COVID-19 reported in Washington state, available at www.livescience.com/coronavirus-spread-washington-nursing-home.html, accessed on April 4, 2020

111 One chart shows New York City's coronavirus cases, deaths, and hospitalizations by age bracket as the city's cases top 20,000, available at www.businessinsider.in/science/news/one-chart-shows-new-york-citys-coronavirus-cases-deaths-and-hospitalizations-by-age-bracket-as-the-citys-cases-top-20000/articleshow/74837730.cms, accessed on April 4, 2020

112 Coronavirus update: Health care worker tests positive after returning from Iran, husband awaiting results, *CBS*, available at https://newyork.cbslocal.com/2020/03/02/first-coronavirus-case-confirmed-in-nyc/, accessed on April 4, 2020

113 He could have seen what was coming: Behind Trump's failure on the virus, *The Economic Times*, available at https://economictimes.indiatimes.com/news/international/world-news/he-could-have-seen-what-was-coming-behind-trumps-failure-on-the-virus/articleshow/75108189.cms?from=mdr, accessed on May 15, 2020

114 The point, *CNN*, available at https://edition.cnn.com/videos/politics/2020/04/07/trump-downplays-coronavirus-mike-pence-depends-cillizza-the-point.cnn, accessed on May 15, 2020

115 Donald Trump ignored 12 warnings issued by CIA about coronavirus, *The Times of India*, available at https://timesofindia.indiatimes.com/world/us/cia-had-issued-12-warnings-to-donald-trump-about-coronavirus-which-he-ignored/articleshow/75443590.cms, accessed on May 15, 2020

116 First coronavirus-related death reported in New York, *CNN*, available at https://edition.cnn.com/2020/03/14/us/nyc-coronavirus-related-death/index.html, accessed on May 17, 2020

117 Gasping for air, delusional and all alone – One COVID-19 survivor's story, *The Print*, available at https://theprint.in/world/gasping-for-air-delusional-and-all-alone-one-covid-19-survivors-story/396531/, accessed on April 10, 2020

118 Cuomo orders most New Yorkers to stay inside – 'we're all under quarantine now', *CNBC*, available at www.cnbc.com/2020/03/20/new-york-gov-cuomo-orders-100per cent-of-non-essential-businesses-to-work-from-home.html, accessed on May 14, 2020

119 Ibid.

120 Coronavirus in New York: Latest updates, *Intelligencer*, available at https://nymag.com/intelligencer/article/new-york-coronavirus-cases-updates.html, accessed on April 4, 2020

121 Why some states became coronavirus hot spots – and others haven't, available at www.vox.com/2020/4/21/21224944/coronavirus-hot-spots-covid-new-york-michigan-florida, accessed on May 15, 2020

122 Statement from the press secretary regarding the President's Coronavirus task force, available at www.whitehouse.gov/briefings-statements/statement-press-secretary-regarding-presidents-coronavirus-task-force/, accessed on May 15, 2020

123 Warning–Level 3, avoid nonessential travel–Widespread ongoing transmission, CDC, available at https://wwwnc.cdc.gov/travel/notices/warning/coronavirus-global, accessed on May 15, 2020

124 Cuomo orders most New Yorkers to stay inside – 'we're all under quarantine now', *CNBC*

125 Coronavirus disease 2019 (COVID-19) situation report – 65, WHO, available at www.who.int/docs/default-source/coronaviruse/situation-reports/20200325-sitrep-65-covid-19.pdf?sfvrsn=ce13061b_2, accessed on May 15, 2020

126 Coronavirus disease 2019 (COVID-19) situation report – 86, WHO

127 Coronavirus disease (COVID-19) situation report – 116, WHO

128 Government of Canada takes action on COVID-19, available at www.canada.ca/en/public-health/services/diseases/2019-novel-coronavirus-infection/canadas-reponse/government-canada-takes-action-covid-19.html, accessed on May 15, 2020

129 Canada: Screening measures implemented in airports January 22, available at www.garda.com/crisis24/news-alerts/308236/canada-screening-measures-implemented-in-airports-january-22, accessed on May 15, 2020

130 Silverstein WK, Stroud L, Cleghorn GE, et al., First imported case of 2019 novel coronavirus in Canada, presenting as mild pneumonia, *Lancet*, available at www.thelancet.com/action/showPdf?pii=S0140-6736%2820%2930370-6, accessed on April 4, 2020

131 First potential coronavirus-related death reported in Ontario, available at https://globalnews.ca/news/6689357/ontario-coronavirus-death/, accessed on April 4, 2020

132 Canadian PM Trudeau's wife tests positive for coronavirus, *BBC*, available at www.bbc.com/news/world-us-canada-51860702, accessed on May 15, 2020

133 Coronavirus disease 2019 (COVID-19) situation report – 86, WHO

134 Coronavirus disease (COVID 19): Outbreak update, available at www.canada.ca/en/public-health/services/diseases/2019-novel-coronavirus-infection.html, accessed on April 4, 2020

135 Coronavirus disease (COVID-19) situation report – 116, WHO

136 First novel coronavirus case reported in Australia, available at https://tass.com/world/1112907, accessed on April 4, 2020

137 Coronavirus: Three cases in NSW and one in Victoria as infection reaches Australia, *The Guardian*, available at www.theguardian.com/science/2020/jan/25/coronavirus-five-people-in-nsw-being-tested-for-deadly-disease, accessed on April 4, 2020

138 Coronavirus: Man evacuated from Diamond Princess becomes first Australian to die of COVID-19, *The Guardian*, available at www.theguardian.com/world/2020/mar/01/dutton-says-extending-travel-ban-not-possible-and-defends-coronavirus-response, accessed on April 4, 2020

139 Coronavirus: First cases of community transmission confirmed in Australia, *The Guardian*, available at www.theguardian.com/world/2020/mar/02/coronavirus-first-cases-of-community-transmission-confirmed-in-australia, accessed on April 4, 2020

140 www.abc.net.au/news/2020-03-19/bridget-wilkins-explains-what-happens-when-you-have-coronavirus/12070370, accessed on April 10, 2020

141 Tom Hanks returns to LA after bout of coronavirus – Media reports, *Reuters*, available at https://in.reuters.com/article/health-coronavirus-hanks/tom-hanks-returns-to-la-after-bout-of-coronavirus-media-reports-idINKBN21F05G, accessed on April 4, 2020

142 https://abcnews.go.com/GMA/News/coronavirus-survivors-symptoms-presented-road-recovery-disease/story?id=69660094, accessed on April 10, 2020

143 COVID-19 student communication, The University of Queensland, available at https://about.uq.edu.au/covid-19-student-communication, accessed on May 17, 2020

144 Coronavirus disease 2019 (COVID-19) situation report – 86, WHO

145 Coronavirus disease (COVID-19) situation report – 116, WHO

5

EMERGING GLOBAL HOTSPOTS OF COVID-19

After decimating China, parts of Europe and the US, the novel coronavirus turned its attention towards the rest of Europe, Asia, Africa and Latin America. In these areas, in the initial few months, the number of cases and deaths were low. However, of late, these areas are facing a rapidly increasing number of cases, and the death toll is climbing rapidly. Previously unaffected areas are rapidly turning into new playgrounds for this deadly virus.

Pandemic in the rest of Europe

Confirmed cases and the number of deaths are increasing daily all over the world. Switzerland, Belgium, Netherlands, Austria, Portugal, Israel, Sweden, Norway, Ireland, Belarus and Denmark have reported more than 20,000 confirmed cases each and many deaths due to COVID-19.[1] However, two countries in particular, Russia and Turkey, deserve special mention. Both countries were able to keep a check on the novel coronavirus in the months of February and March of 2020. There were no huge spikes in the number of cases. However, in the months of April and May, the number of cases shot up rapidly, causing significant death and debility.

Russia

On January 31, 2020, Russia reported the presence of coronavirus when two Chinese tourists tested positive for the virus in Tyumen, Siberia and Chita, Russian Far East.[2] Both patients recovered well from their illness. In response to it, in the first week of February 2020, Russia started checking the body temperatures of people attending events and restricted entry from China. In February, few cases were reported. According to the WHO's situation report, dated March 15, 2020,

DOI: 10.4324/9781003345091-5

there were 34 confirmed cases in Russia without any mortality.[3] Many of the new cases were contracted in Italy. Meanwhile, seven Russian passengers from the cruise ship Diamond Princess who were placed in a two-week quarantine in Russia's Far East were discharged.[4] According to the *Moscow Times*, a Russian national who was quarantined in a Moscow hospital after returning from northern Italy tested positive for coronavirus. The man, 29-year-old Moscow resident D. B., became the first-known Russian national to test positive for coronavirus while in the country.[5]

Gradually, the infection spread far and wide in Russia. Moscow became the most-affected place.[6] On March 19, 2020, a 79-year-old Russian woman died in a Moscow hospital after testing positive for the coronavirus. It was the first coronavirus-related death in the country.[7] According to the WHO's situation report dated March 31, the total number of confirmed cases increased to 1837, with nine deaths.[8] The number of cases steadily increased. There were no sudden spikes in March 2020. The high dispersion and low mobility of Russian people were among the reasons for the relatively low coronavirus incidence in Russia, according to the analytical report titled 'The influence of the COVID-19 coronavirus on situation in Russian healthcare', prepared by the Ministry of Health's Central Research Institute of Healthcare Organization and Informatics.[9] The low incoming tourist flow, low tourism activity of Russians and low development of internal tourism, especially during winter, were also identified among the important reasons for the low incidence of COVID-19 cases in Russia.[10]

To prevent further spread of infection, the Russian government implemented several measures, including the imposition of quarantine, banning the export of medical masks,[11] the cancellation of large-scale events in schools, the closure of all sports facilities (including swimming pools and fitness clubs), banning all gatherings in public places and closing the national border.[12] Widespread testing for coronavirus was carried out for the early detection and management of cases. In February and early March, additional measures restricting access to foreigners – first Chinese people, then Iranians, and finally people of other nationalities – were adopted by the authorities, but the country allowed Russians to return from holidays in Europe and elsewhere, many of whom were carriers of the virus. On March 24, 2020, President Vladimir Putin visited one of Moscow's largest hospitals, Kommunarka, where patients were being treated. According to critics, after this, he failed to implement the strong measures required to contain such a deadly pandemic.[13] However, Moscow Mayor Sergei Sobyanin ordered all residents to self-isolate at home.[14] The government declared a non-working period, which lasted up to May 11, 2020, after being extended twice.

In spite of all of these efforts, the number of cases steadily increased, and by April 15, 2020, the number of confirmed cases had increased to 24,490, and death toll had increased to 198.[15] In the second half of April 2020, on average, the daily increase was between 5000 and 6000 cases. As a result, the total number of cases by April 30, 2020, surpassed that of China. According to the WHO's situation report dated April 30, 2020, the total number of confirmed cases had increased to 106,498, and the number of deaths had risen to 1073.[16] On April 30, Prime

Minister Mikhail Mishustin tested positive for coronavirus.[17] Mr Mishustin took the opportunity to urge all Russians to take coronavirus seriously and to stay at home.

In May 2020, the situation changed dramatically, and by May 12, Russia had 232,243 confirmed cases of coronavirus, the second highest toll in the world after the US. The death toll had climbed to 2116.[18] In the previous 24 hours, the country reported 10,899 infections, the tenth consecutive day that that number had been above 10,000. Among the infected was President Vladimir Putin's spokesperson, Dmitry Peskov.[19] Mr Putin announced the end of six weeks of non-working days, and Russians began returning to work on Tuesday morning.[20] Moscow became the epicentre of Russia's outbreak, accounting for more than half of the country's cases and more than half of its death toll.[21]

As per the WHO's situation report dated May 15, 2020, there were 262,843 confirmed cases, with 2418 deaths (Figure 5.1a and Figure 5.1b).[22] In Russia, the number of confirmed cases is very high, but the death toll is relatively low. There are various possible reasons for the high number of cases. First, the testing rate in Russia is one of the highest in the world. So the country is able to diagnose the maximum number of cases. Due to early diagnosis and the early initiation of treatment, the death rate has remained low. However, there are other factors

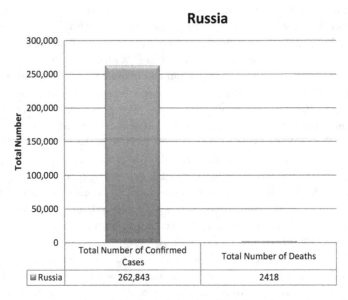

FIGURE 5.1a Total Number of Confirmed Cases and Deaths in Russia Due to COVID-19 by May 15, 2020

Source: WHO.

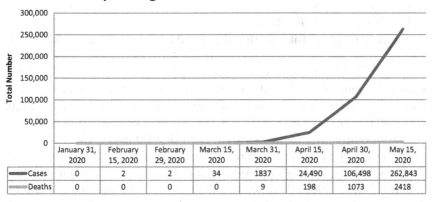

Rapid Progression of Cases and Deaths in Russia

	January 31, 2020	February 15, 2020	February 29, 2020	March 15, 2020	March 31, 2020	April 15, 2020	April 30, 2020	May 15, 2020
Cases	0	2	2	34	1837	24,490	106,498	262,843
Deaths	0	0	0	0	9	198	1073	2418

FIGURE 5.1b Rapid Progression in the Total Number of Confirmed Cases and Deaths Due to COVID-19 in Russia by May 15, 2020

Source: WHO.

for the high number of cases. International traffic continued until the middle of March. In the initial days, people disregarded lockdown and moved around carelessly, which led to a further tightening of the lockdown.[23] However, the pandemic is far from over in Russia, and the situation may further worsen with the easing of the lockdown.

Turkey

The novel coronavirus knocked at the door of Turkey much later than it did in many countries. The first case was reported on March 11, 2020.[24] Health Minister Fahrettin Koca said in a news conference that the person contracted the virus while travelling to Europe. The minister urged people not to be alarmed by the diagnosis. He said, 'The coronavirus is not strong enough to break through the measures taken by Turkey'.[25] Coronavirus proved him wrong. On March 13, Health Minister Koca announced that the total number of confirmed cases had increased to five.[26]

Turkey had implemented several measures even before the cases were confirmed, but unfortunately, they were not enough. In the month of January 2020, the Ministry of Health set up the Coronavirus Scientific Advisory Board.[27] Thermal cameras were first installed at airports to prevent the entry of infected passengers from abroad, and flights to and from Italy, South Korea, Iran and Iraq were later cancelled. In early February, flights from China were cancelled. On February 23, border crossings with Iran were shut down.[28] Iraqi borders have since also been temporarily closed. Disinfection and sanitation drives were implemented to halt the virus. In March, Friday prayers gatherings were banned. Later, cafés, gyms and cinema halls were closed.

All these measures failed to stop the spread of the virus, and on March 17, Turkey confirmed the first death due to COVID-19. Health Minister Koca told a press conference that an 89-year-old patient died after contracting the virus from someone who had contact with China.[29] The next day, another patient died of COVID-19. The WHO situation report dated March 31, 2020, reported 10,827 confirmed cases and 168 deaths.[30] Further, all 81 provinces reported positive cases in the first week of April 2020, and Istanbul accounted for 60% of the total cases across the country.[31] The health minister announced that 601 health personnel had been counted as among those infected and confirmed the death of Professor Cemil Taşçıoğlu, who was among the team that treated the first patient diagnosed with the coronavirus. Taşçıoğlu's death marked the first casualty of the Turkish medical community in the fight against coronavirus.[32]

Death and debility continued without any sign of stoppage. The number of confirmed cases increased at an alarming rate in the month of April. Face masks were mandatory on public transport, in markets and in other communal spaces. Thirty-one cities were closed to all but essential traffic. Dr F Koca

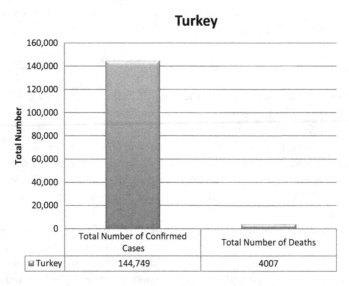

FIGURE 5.2a Total Number of Confirmed Cases and Deaths in Turkey Due to COVID-19 by May 15, 2020

Source: WHO.

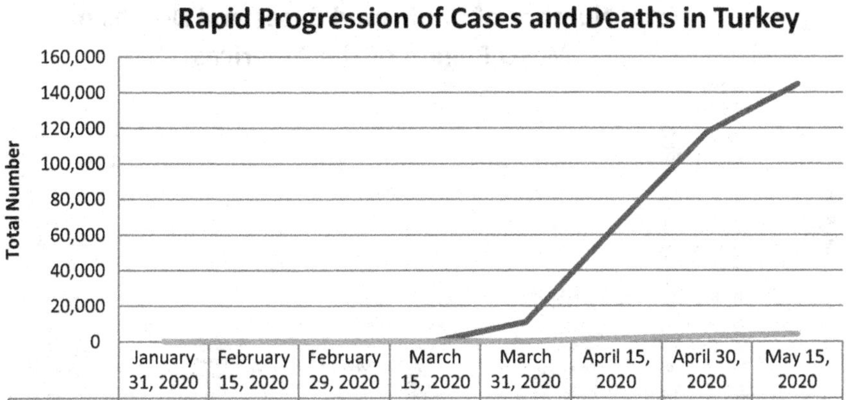

Rapid Progression of Cases and Deaths in Turkey

	January 31, 2020	February 15, 2020	February 29, 2020	March 15, 2020	March 31, 2020	April 15, 2020	April 30, 2020	May 15, 2020
Cases	0	0	0	5	10,827	65,111	117,589	144,749
Deaths	0	0	0	0	168	1403	3081	4007

FIGURE 5.2b   Rapid Progression in the Total Number of Confirmed Cases and Deaths Due to COVID-19 in Turkey by May 15, 2020

Source: WHO.

urged people to 'stay at home', saying the virus 'draws its power from contact'.[33] President Recep Tayyip Erdogan asked people to practise social distancing and stay 'three paces' from one another. However, complete lockdown was not imposed.[34] The WHO situation report dated April 30 indicated that there were 117,589 confirmed cases and 3081 deaths.[35] The condition in Turkey is getting worse with every passing day, and it has now the maximum number of cases in the Middle East and is among the top ten worst-affected countries in the world. According to the WHO's situation report dated May 15, 2020, the total number of confirmed cases was 144,749, with 4007 deaths (Figure 5.2a and Figure 5.2b).[36] And the pandemic is not showing any signs of slowing down yet.

Pandemic in the Region of the Americas

In the Americas (excluding the US and Canada), Brazil, Chile, Ecuador, Peru, Colombia and Mexico (Figure 5.3) have reported several thousand cases each.[37] Latin America has the potential to turn into the next hotspot in the world. Poverty and a lack of adequate healthcare infrastructure in Latin America may have disastrous consequences. Brazil deserves a special mention, because the number of confirmed cases and the death toll have increased significantly, making Brazil one of the worst-affected countries in this region and the world.

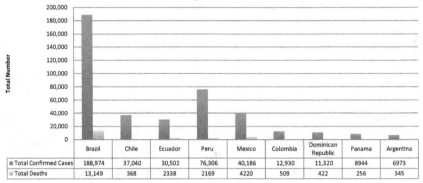

FIGURE 5.3 Total Number of Confirmed Cases and Deaths in Region of the Americas by May 15, 2020

Source: WHO.

Brazil

The COVID-19 pandemic reached Latin America later than it did in other continents. On February 25, 2020, the first case of COVID-19 was reported in Brazil when a man tested positive for coronavirus in São Paulo.[38] He had returned from Lombardy. The second case was reported soon after.

On March 12, 2020, Fabio Wajngarten, the press secretary for Brazilian President Jair Bolsonaro, tested positive for coronavirus.[39] Bolsonaro's health was also being monitored. It happened just days after Wajngarten met US President Donald Trump in Florida. The Brazilian press secretary attended the dinner that Trump hosted at his Mar-a-Lago resort in Florida, took a photo with the US president and later stood close to Trump.

The WHO's situation report dated March 19, 2020, reported the first COVID-19 death in Brazil.[40] However, the BBC reported that the death happened on March 16, 2020.[41] Meanwhile, scientists at the Oswaldo Cruz Foundation examined cases of patients taken to hospital with respiratory problems. According to them, molecular tests suggested one patient who died in Rio de Janeiro between January 19 and 25 had COVID-19. The scientists said that their research suggested that the virus was being spread from person to person in Brazil in early February, weeks before the country's popular Carnival street parties kicked off.[42] By March 31, there were 4256 confirmed cases and 136 deaths.[43]

The virus steadily spread from one end of the country to another. Large cities such as São Paulo and Rio de Janeiro became the main hotspots,[44] but infection gradually spread inland into smaller cities, with inadequate provisions for intensive care beds and ventilators. The *Guardian* reported on April 10, 2020, that a Yanomami teenager had reportedly died after contracting COVID-19,[45] further fuelling fears over the disease's potential to decimate Indigenous communities in the Amazon. The website Amazônia Real reported that the village's 70 or so members had been isolated, as had the victim's parents, five health workers and a local pilot. The death pointed out the coronavirus's potential to wreak havoc on Indigenous communities across South America.[46]

The cases have been increasing rapidly in April and May 2020. The WHO's situation report dated April 30, 2020, reported 71,886 confirmed cases and 5017 deaths.[47] And by May 15, 2020, the number of cases had risen to 188,974, and the death toll had climbed to 13,149 (Figure 5.4a and Figure 5.4b).[48] Brazil is now among the top ten worst-affected countries in the world. The situation in Manaus deteriorated rapidly.[49] Deaths from the coronavirus outbreak piled up so fast in the Amazon rainforest's biggest city that the main cemetery had been burying five coffins at a time in collective graves. Manaus, the capital of Amazonas state, was the first in Brazil to run out of ICU beds, but officials warned that several other cities were close behind. In Rio de Janeiro, cemeteries have accelerated the construction of above-ground vaults to entomb a wave of deceased patients.[50] According to the *Lancet*, the doubling of the rate of deaths is estimated at only five days, and a recent study by Imperial College (London, UK), which analyzed the active transmission rate of COVID-19 in 48 countries, showed that Brazil is the country with the highest rate of transmission (R_0 of 2·81).[51]

The WHO says the Americas are currently at the centre of the pandemic.[52] The outbreak is expected to accelerate over the coming weeks, and there are fears that the

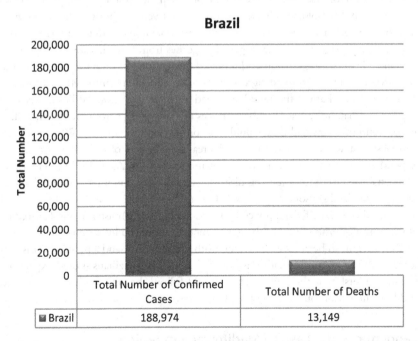

FIGURE 5.4a   Total Number of Confirmed Cases and Deaths in Brazil Due to COVID-19 by May 15, 2020

Source: WHO.

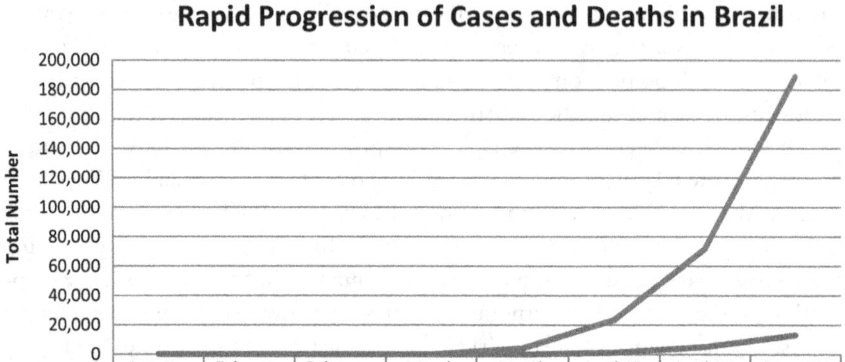

Rapid Progression of Cases and Deaths in Brazil

	January 31, 2020	February 15, 2020	February 29, 2020	March 15, 2020	March 31, 2020	April 15, 2020	April 30, 2020	May 15, 2020
Cases	0	0	1	121	4256	23,430	71,886	188,974
Deaths	0	0	0	0	136	1328	5017	13,149

FIGURE 5.4b   Rapid Progression in the Total Number of Confirmed Cases and Deaths Due to COVID-19 in Brazil by May 15, 2020

Source: WHO.

pandemic could overwhelm Brazil's health system. Brazil's total number of confirmed cases and death toll is second only to that of the US in the Western Hemisphere. However, President Jair Bolsonaro has repeatedly downplayed the threat of the coronavirus and criticized governors and mayors for adopting strict restrictions to curb its spread.[53] The pandemic triggered various types of responses from federal, state and local governments. Most state governors had imposed quarantine to prevent the spread of the virus. Many state governors disagreed with the president and imposed lockdown in their own states. Earlier, the president issued a decree that classified businesses such as gyms and hairdressers as 'essential' services that are exempt from lockdowns.[54] But at least ten governors said they would not comply with the order.[55] When asked by journalists last week about the rapidly increasing numbers of COVID-19 cases, he responded: 'So what? What do you want me to do?' On April 16, Luiz Henrique Mandetta, the health minister, was dismissed after a television interview in which he strongly criticized Bolsonaro's actions.[56] Further, on April 24, following the removal of the head of Brazil's federal police by Bolsonaro, Justice Minister Sérgio Moro, one of the most powerful figures of the government, announced his resignation.[57]

Brazil seems to be at its worst facing both a health crisis and a political crisis. If a solution to the political situation is not found quickly, Brazilians should expect the worst. Leadership at the highest level should follow the right path to avoid crisis and lead the nation out of one of its darkest days.

Pandemic in the Eastern Mediterranean region

Besides Iran, in the Eastern Mediterranean region, Saudi Arabia, Pakistan, Qatar, Kuwait and the United Arab Emirates (UAE) have also reported more than 10,000

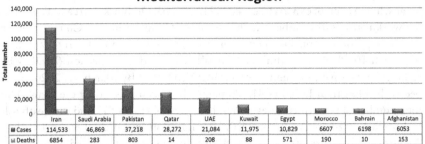

Total Number of Confirmed Cases and Deaths in Eastern Mediterranean Region

	Iran	Saudi Arabia	Pakistan	Qatar	UAE	Kuwait	Egypt	Morocco	Bahrain	Afghanistan
Cases	114,533	46,869	37,218	28,272	21,084	11,975	10,829	6607	6198	6053
Deaths	6854	283	803	14	208	88	571	190	10	153

FIGURE 5.5   Total Number of Confirmed Cases and Deaths in the Eastern Mediterranean Region by May 15, 2020

Source: WHO.

cases each (Figure 5.5). By May 15, 2020, in the Eastern Mediterranean region, there were 305,189 confirmed cases and 9558 deaths.[58] Iran deserves a special mention here, because it reported the maximum number of cases with high death toll, making it the worst-affected country in this region.[59]

Iran

The raging COVID-19 pandemic did not spare the usually dry and hot Middle East, and Iran quickly became the epicentre of a pandemic in the region. According to the *New Yorker*, the outbreak appears to have started in Qom,[60] the conservative city of Shiite seminaries, about two hours from Tehran. The first mention of the presence of coronavirus in Iran was announced by the government when it reported two deaths in the city on February 19, 2020. The carrier of the virus may have been a merchant who travelled between Qom and Wuhan, China. The outbreak is thus estimated to have started between three and six weeks earlier.[61]

Within eight days COVID-19 spread to 24 of the country's 31 provinces. From Iran, coronavirus reached several countries, including Azerbaijan, Afghanistan, Bahrain, Canada, Georgia, Iraq, Kuwait, Lebanon, Oman, Pakistan and the UAE. Many of these cases are directly linked to visits to Qom.[62]

On February 24, 2020, Iran's deputy minister of health, Dr Iraj Harirchi, reported that 12 people had died and up to 61 had been infected with the new coronavirus.[63] Later, multiple government ministers and officials, including several members of Parliament, were infected with coronavirus.[64] According to media reports, at least 12 Iranian politicians and officials, both sitting and former, have now died of the illness, and 13 more have been infected and are either in quarantine or being treated.[65] Dr Harirchi was pale and drenched in sweat during a press conference as he told reporters that the Islamic Republic had 'almost stabilized' the country's outbreak of coronavirus. He mopped his brow often during the press conference and later confirmed that he had contacted the coronavirus and was under quarantine.[66]

The WHO situation report dated March 1, 2020, reported 593 cases and 43 deaths,[67] which climbed to 12,729 cases and 608 deaths as of March 15, 2020.[68] Meanwhile, a team of experts from the WHO, Global Outbreak Alert and Response Network partners, the Robert Koch Institute in Germany and the Chinese Center for Disease Control concluded a technical support mission on COVID-19 to Iran on March 10, 2020. The Ministry of Health and Medical Education (MOHME) launched a national campaign to control COVID-19.[69] By March 31, 2020, 2757 people had lost their lives and 41,495 people had been affected by the virus. The total number of confirmed cases in Iran had increased to 74,877 and the death toll had reached 4683 by April 15, 2020. The infection spread throughout Iran like wildfire, and on April 30, 2020, the WHO situation report revealed 93,657 confirmed cases and 5957 deaths.[70]

The Iranian government, in response to coronavirus infection, took several steps to mitigate the outbreak. The government cancelled public events and Friday prayers and closed bazaars, shopping malls, universities, schools and holy shrines. It also took several steps to help poor families and businesses. However, it did not place entire cities under quarantine. Later, the government announced travel bans between cities to limit further spread.

According to *Al Jazeera*, Iran's supreme leader, Ayatollah Al Khamenei, refused American assistance to fight the pandemic, citing a conspiracy theory claiming that it could have been manufactured by the US.[71] He further said that it could be a biological attack.

Many scholars have reported that the total number of cases reported by the government to be too low and the death toll to be much higher. The government has also been accused of coverups, censorship and mismanagement. The government has been accused of having squandered multiple opportunities to respond to the pandemic, first by failing to address it squarely when the coronavirus entered Iran and later on by neglecting expertise-based proposals to contain its spread. However, the WHO has supported the figures reported by the government. On March 17, WHO Regional Emergency Director Rick Brennan said the number of cases reported in Iran could represent only about 20% of the real numbers, because testing was restricted to severe cases, as is the case even in some wealthy European countries.[72]

On April 15, 2020, the BBC reported on a video of a mortuary worker at a cemetery in Qom surrounded by dozens of bodies.[73] The man who filmed it was later arrested, and the authorities informed the public that 'all bodies were being treated with respect and in keeping with Islam'. The BBC also reported that a doctor from the Mazandaran region of northern Iran told the BBC that public safety experts had been dispatched to monitor the process of enshrouding, burying and covering the graves with lime.[74]

Among all these claims and counterclaims and in spite of all efforts by the government, the cases continued to rise, and by May 15, 2020, the number of cases had increased to 114,533, and the death toll had climbed to 6854 (Figure 5.6a and Figure 5.6b).[75] Iran at present is trying its best to contain the pandemic and limit the death toll.

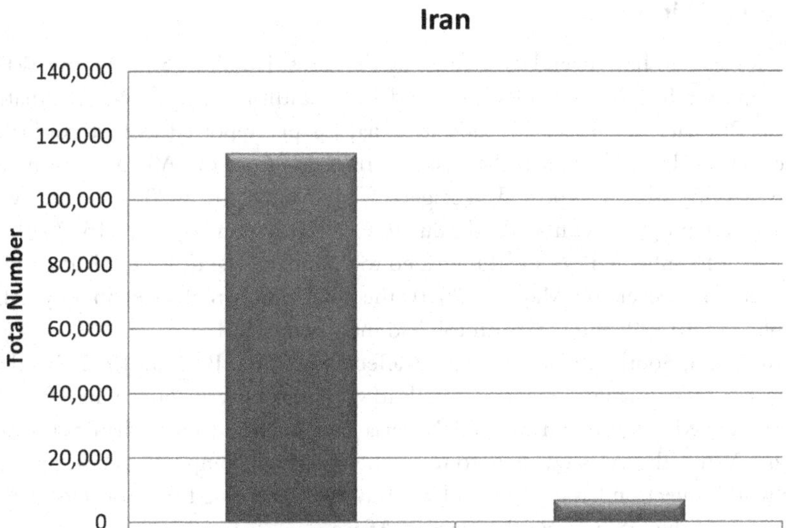

Iran

	Total Number of Confirmed Cases	Total Number of Deaths
■ Iran	114,533	6854

FIGURE 5.6a   Total Number of Confirmed Cases and Deaths in Iran Due to COVID-19 by May 15, 2020

Source: WHO.

Rapid Progression of Cases and Deaths in Iran

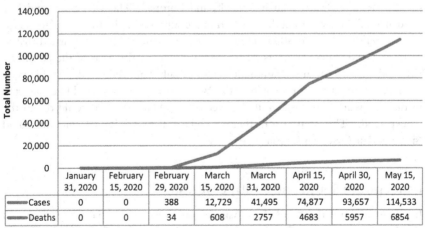

	January 31, 2020	February 15, 2020	February 29, 2020	March 15, 2020	March 31, 2020	April 15, 2020	April 30, 2020	May 15, 2020
Cases	0	0	388	12,729	41,495	74,877	93,657	114,533
Deaths	0	0	34	608	2757	4683	5957	6854

FIGURE 5.6b   Rapid Progression in the Total Number of Confirmed Cases and Deaths Due to COVID-19 in Iran by May 15, 2020

Source: WHO.

Pandemic in Africa

Africa seems to have been less affected by the virus. By May 15, 2020, the WHO had reported 54,190 confirmed cases and 1623 deaths in Africa.[76] WHO situation report 26, dated February 15, indicated that Egypt[77] reported its first confirmed case of COVID-19.[78] This is the first reported case from the African continent.[79] The second case was reported in Algeria.[80] By March 31, 2020, 656 cases were confirmed in Egypt, with 41 deaths due to COVID-19, and by April 15, 2020, the number of confirmed cases had increased to 2350, and the death toll had climbed to 178.[81] However, by May 15, 2020, the total number of cases in Egypt had climbed to 10,829, and the death toll had increased to 571.[82]

In Africa, South Africa is the worst-affected country. By May 15, 2020, there were 12,739 confirmed cases and 238 deaths.[83] Almost every country in Africa has been affected. Algeria, Ghana and Nigeria have reported more than 5000 cases each.[84] While the virus was slow to reach the continent compared to other parts of the world, infection has grown rapidly.[85] Travellers returning from hotspots of Asia, Europe and the US brought the virus to Africa.

COVID-19 is capable of causing massive socioeconomic devastation. The WHO is working with governments across Africa to scale up surveillance, testing, isolation, case management, contact tracing, infection prevention and control, risk communication, community engagement and laboratory capacity.[86] In many parts of the world, the number of cases has started falling, but in Africa, the number of cases is rising, and many believe that the continent may turn out to be the next hotspot.[87] A lack of testing kits and lack of adequate infrastructure are the main bottlenecks in the fight against COVID-19.[88]

As reported in the *Guardian*, nearly a quarter of a billion people across 47 African countries will catch coronavirus over the next year, but the result will be fewer severe cases and deaths than in the US and Europe.[89] The young age and low incidence of obesity among the population are supposed to be the main factors for fewer severe cases and deaths.[90] However, recently, the WHO has warned that as many as 190,000 people across Africa could die in the first year of the coronavirus pandemic if crucial containment measures fail. The study finds that between 29 million and 44 million people in the WHO African region could get infected in the first year of the pandemic. Between 83,000 and 190,000 could die in the same period.[91] The WHO has further warned that the COVID-19 pandemic could 'smoulder' in Africa for several years.[92]

Pandemic in South and South-East Asia

In the South and the South-East Asia region India, Indonesia, Bangladesh and Thailand have reported a few thousand cases each.[93] By May 15, 2020, 18,863 cases and 283 deaths had been reported in Bangladesh. Similarly, Indonesia has 16,006 confirmed cases with 1043 deaths. In Thailand, the total number of cases is 3025, and the death toll is 56. India is the worst-affected country in this region. India has the potential to become the next epicentre of the epidemic. The virus may spread like wildfire among its huge population.

India

The WHO situation report dated January 30, 2020,[94] confirmed the first case in India, and by February 1, 2020, there were a total of three cases, and all had a history of travel to China. All three cases were detected in the southern state of Kerala. Meanwhile, India started airlifting stranded Indian students from Wuhan, China. The first airlift was conducted on February 1, 2020, and the final airlift was conducted on February 27, 2020.

In first week of March 2020, two more cases were reported. A 45-year-old man who had returned from Italy and a 24-year-old man who had returned from UAE tested positive for COVID-19. Besides this, an Italian citizen was found to be positive for COVID-19 in Jaipur. Contact tracing was carried out promptly, and several contacts were found to have symptoms of COVID-19. Sporadic case reports started appearing from different parts of India, including Ghaziabad, West Delhi and Ladakh. The majority of these cases had a history of travel to various COVID-19-affected countries.[95] By this time, the Indian government had quarantined almost all the contacts of these cases to prevent further spread. By the middle of March 2020, cases had been reported from various states of India, including Karnataka, Uttar Pradesh, Maharashtra and Telengana.

Surprisingly, even by end of March 31, 2020, the total number of cases was 1071, and the death toll was 29, which is very low compared to the population. By April 15, 2020, the total number of confirmed cases was 11,439, and the death toll had climbed to 377. By May 15, 2020, the total number of confirmed case in India had increased to 81,970, and the death toll had climbed to 2649 (Figure 5.7a and Figure 5.7b).[96] Although the cases and the death toll have started increasing lately, they have remained low given the country's population density. This can be attributed to low testing rates in India, due to a shortage of test kits. But the low number of deaths cannot be attributed to a lack of testing. Every critical patient admitted to hospital is tested for COVID-19. There is hardly any overwhelming of the inadequate healthcare infrastructure in India. The exact reason behind this is difficult to know at this moment and will need to be researched after the pandemic is over.

The Indian government, unlike many other countries, took the evolving situation seriously and implemented various measures to contain the infection. The few cases in India in the month of February and early March gave the government enough time to improve the healthcare infrastructure to tackle the surge of cases. As per the report published in the *Hindu*,[97] dated April 10, 2020, Health Minister

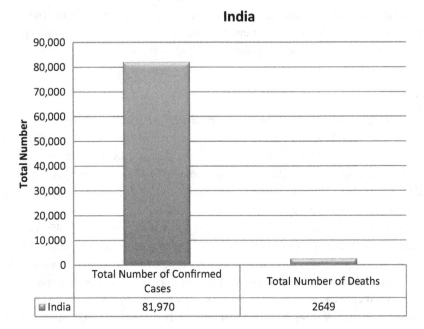

FIGURE 5.7a Total Number of Confirmed Cases and Deaths in India Due to COVID-19 by May 15, 2020

Source: WHO.

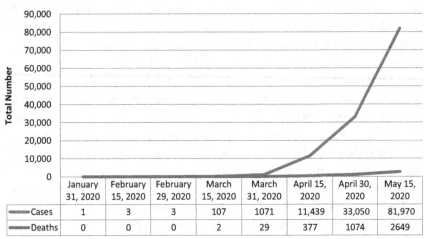

FIGURE 5.7b Rapid Progression in the Total Number of Confirmed Cases and Deaths Due to COVID-19 in India by May 15, 2020

Source: WHO.

Harsh Vardhan said, 'were matters to deteriorate, India had enough infrastructure to tide over the crisis'. He added,

> We have 500 dedicated COVID-19 hospitals, 200,000 beds across hospitals and 50,000 ICU beds. We have ordered everything in plenty. In the last two months, we have ensured adequate testing facilities. We started with one lab and today we have over 200. Over 150 are public labs.

The government also had enough time to study the development of the pandemic in various other countries and implement effective measures. India implemented the provisions of the Epidemic Diseases Act, 1897. All education institutions were closed, public gatherings of all types were banned and all nonessential commercial establishments were shut down. India suspended all tourist visas.

On March 22, 2020, India observed a voluntary public curfew at the instance of Prime Minister Narendra Modi. Furthermore, on March 24, 2020, Mr Modi ordered an unprecedented nationwide lockdown for 21 days, affecting all the 1.3 billion people in India.[98] On April 14, 2020, it was further extended to May 3, 2020. However, the cases were still increasing, and as a result, the lockdown was extended until May 17, 2020,[99] which was further extended until May 31, 2020.[100]

The lockdown gave just four hours for the people to prepare, and as a result, panic buying started in the markets, and social-distancing norms were violated. Many migrant workers in various cities of India were stranded without food and money, and as a result, a mass exodus started. Thousands of stranded workers wanted to return home to their villages and in the absence of transport were forced to walk hundreds of kilometres under the scorching sun. The pitiable condition of a huge section of the Indian population and the lack of preparedness of the government were brought out in the open. Many workers lost their jobs, and many were forced below the poverty line as a result. The Indian government had to face criticism for its ill-preparedness.[101]

Meanwhile, the executive director of the WHO, Michael Ryan, said India, the world's second-most-populous country, has a tremendous capacity to deal with the coronavirus outbreak because it has the experience of eradicating small pox and polio through targeted public intervention.[102] The Oxford COVID-19 Government Response Tracker, based on data from 73 countries, reported that the government of India had responded more stringently than any other country in tackling the COVID-19 pandemic. The tracker is based on indicators such as school closures, travel bans and measures such as emergency investments in health-care, fiscal measures and investments in vaccines. It noted the Indian government's swift action, emergency policymaking, emergency investment in healthcare, fiscal measures, investments in vaccine research and active response to the situation, and it scored India with a '100' for its strictness.[103]

Under economic compulsion,[104] the Indian government started relaxing some strictures in the last week of April 2020.[105] As a result, people started moving about, and cars and two wheelers started moving in the streets.[106] Besides this, in India, there are many who live on the footpaths, and for these homeless people, stay-at-home orders are meaningless. In the overcrowded cities where five to six people are forced to live in a single room, social distancing, which should be called physical distancing, becomes meaningless.

The epidemic is still evolving in India, and it is difficult to predict what will happen there. However, warm weather, Bacillus Calmette-Guérin (BCG) inoculation and a fairly good public health infrastructure are the possible ways that it can check its death toll. India has recently eradicated polio, and the public health machinery is intact, unlike many developed nations. Case detection, surveillance, contact tracing, travel bans, social distancing and strict lockdown have to date and to some extent helped in containing the epidemic and flattening the curve. However, the end of the pandemic is not yet in sight, and with a rapidly increasing number of cases and a climbing death toll, it seems the road ahead will be rough for India.

Pandemic in the Western Pacific region

As reported by the WHO in the Western Pacific region (excluding China, South Korea and Australia), a significant number of cases have been detected in Singapore, Philippines, Japan and Malaysia.[107] The number of cases in Singapore was low in the early part of pandemic. In March 2020, WHO chief Tedros Adhanom Ghebreyesus had praised Singapore's 'all-government approach' in the containment of COVID-19.[108] However, by May 15, 2020, the number of cases in Singapore increased to 26,098, with 21 deaths. Similarly, Japan reported 16,193 cases and 710 deaths. Philippines reported 11,876 cases and 790 deaths. Malaysia reported 6819 cases and 112 deaths. New Zealand confirmed 1148 cases and 21 deaths.[109]

Cruise ships: pandemic in floating apartments

Historically, pandemics have been severe among urban populations. Densely populated cities with unhygienic conditions are fertile playgrounds for biological foes. Cruise ships, which are equivalent to densely populated cities with cramped apartments and intermingling among passengers and crewmembers from around the world, serve as petri dishes for viruses. Movement from port to port further adds to the problem. As a result, the pandemic rapidly spreads through cruise ships, affecting a significant proportion of passengers in a closed environment.

Diamond Princess

During February and March 2020, COVID-19 outbreaks associated with cruise ship voyages caused more than 800 laboratory-confirmed cases among passengers and crew, along with ten deaths.[110] The Diamond Princess was affected the most. The Diamond Princess departed on its fatal journey from Yokohama on January 20, 2020, with about 3700 passengers and crew.[111] On January 25, a passenger departed the ship with symptoms of coronavirus infection at Hong Kong.

Unfortunately, the passenger tested positive for COVID-19. It was the beginning of the misery. On February 3, 2020, the Diamond Princess returned to Japan after making six stops in three countries, and the ship was promptly quarantined. On February 5, passengers were quarantined in their cabins, but the crew could not be isolated in their cabins.

As of February 8, 64 individuals were found to have been infected among passengers and crew members. All passengers testing positive disembarked and were admitted to hospitals in the Yokohama area for medical care. Close contacts of the infected passengers were asked to remain in quarantine for 14 days from their last contact with a confirmed case. The virus affected 454 people in the Diamond Princess by February 18, 2020. In contrast, the total number of cases in Singapore was only 77 at the same time. By February 21, 2020, the number of cases in Diamond Princess rose to 634.[112]

However, the quarantine was criticized by many, because it was found ineffective at stopping the virus. Dr Anthony Fauci, director of the National Institute of Allergy and Infectious Disease in the US, said, 'I'd like to sugar-coat it and try to be diplomatic about it, but it failed. . . . People were getting infected on that ship. Something went awry'.[113] The cruise ship seemed to have turned into a virus incubator rather than a quarantine. Kentaro Iwata, professor at the infectious diseases division of Japan's Kobe University, described the situation on board as 'completely inadequate in terms of infection control'. He added that the quarantine measures he witnessed failed to separate the infected from the healthy. There was no demarcation between green zone (no infection) and red zone (infected area).[114]

During February 16–23, 2020, nearly 1000 people were repatriated by air to their home countries, including 329 people who returned to the United States and entered quarantine or isolation.[115] Among 3711 Diamond Princess passengers and crew, 712 (19.2%) had tested positive. Of these, 331 (46.5%) were asymptomatic at the time of testing. Among 381 symptomatic patients, 37 (9.7%) required intensive care and nine (1.3%) died.[116] There was utter confusion and a lack of information on the ship. 'There wasn't much information', passenger M. I. told the *New York Times* on February 5, 2020.[117] Many were confined to their rooms, which did not have a balcony or a window. Many passengers started flashing messages that there was not enough medicine onboard and thanked the media for portraying their plight. Some Indian crew members requested the Indian government to rescue them from the ill-fated ship. 'We all are really scared and tense', said S. T., a worker who talked to CNN.[118] The joyride ended in disaster for many.

Grand Princess

The voyage of the Grand Princess, which sailed roundtrip from San Francisco, California, making four stops in Mexico during February 11 to February 21, 2020, also ended in disaster.[119] On March 4, a physician in California reported that a passenger tested positive for coronavirus. The CDC notified the cruise ship and a further 20 people were found to be infected. On March 5, 2020, 21 people, two passengers and 19 crew members had positive test results.[120] Passengers and symptomatic crew members were asked to self-quarantine in their cabins. On March 8, 2020, the ship docked at Oakland, California, and the passengers were put into quarantine for 14 days and hospitalized if indicated.

Besides these two cruise ships, there were other ocean cruise ships and Nile river cruise ships where the virus affected many passengers. The close environment, crew transfer and contact between passengers of different countries allowed the virus to spread.

★★★

The pandemic is spreading rapidly with no end in sight. The number of cases and death tolls are mounting with every passing day. The unaffected areas are rapidly turning into new playgrounds for this deadly virus. There is no vaccine and no therapeutic measures to control the spread. David Nabarro, the WHO's special envoy for COVID-19, has said that the coronavirus is not going to go away, so we must learn to live with it.[121]

Until an effective vaccine becomes available, non-therapeutic measures like lockdown, travel bans and school and college closures will play important roles in controlling the pandemic. However, with the easing of lockdowns in several countries, people have started coming out of their homes. The more people move around, the greater the chance of catching the infection. In such a scenario, it becomes crucial to learn how to recognize a case of COVID-19 and take effective precautionary measures to prevent infection. Besides prevention of infection, it is extremely important to take care of the patients and reduce death and debility from COVID-19. Early and correct diagnoses, based on clinical features and appropriate laboratory tests followed by effective treatment, will go a long way in winning the war against COVID-19.

Notes

1 Coronavirus disease (COVID-19) situation report – 116, WHO, available at www. who.int/docs/default-source/coronaviruse/situation-reports/20200515-covid-19-sitrep-116.pdf?sfvrsn=8dd60956_2, accessed on May 16, 2020

2 Update 1-Russia discharges second Chinese national to recover from coronavirus, *Reuters*, available at https://ru.reuters.com/article/idUKL8N2AC16V, accessed on May 15, 2020

3 Coronavirus disease 2019 (COVID-19) situation report – 55, WHO, available at www.who.int/docs/default-source/coronaviruse/situation-reports/20200315-sitrep-55-covid-19.pdf?sfvrsn=33daa5cb_8, accessed on May 13, 2020

4 Coronavirus in Russia: The latest news, *The Moscow Times*, available at www.themoscowtimes.com/2020/05/13/coronavirus-in-russia-the-latest-news-may-13-a69117, accessed on May 13, 2020

5 Ibid.

6 Global report: Russia becomes Europe's coronavirus hotspot, *The Guardian*, available at www.theguardian.com/world/2020/may/07/global-report-france-to-ease-lockdown-as-russia-becomes-coronavirus-hotspot, accessed on May 15, 2020

7 Elderly woman with coronavirus dies in Moscow, *The Guardian*, available at https://tass.com/society/1132327, accessed on May 15, 2020

8   Coronavirus disease 2019 (COVID-19) situation report – 71, WHO, available at www.who.int/docs/default-source/coronaviruse/situation-reports/20200331-sitrep-71-covid-19.pdf?sfvrsn=4360e92b_8, accessed on May 15, 2020

9   Health Ministry names reasons behind low COVID-19 incidence in Russia, *TASS*, available at https://tass.com/society/1151271, accessed on May 13, 2020

10 Ibid.

11 Russian government restricts exports of face masks, other medical goods till June 1, available at https://tass.com/economy/1126373, accessed on May 15, 2020

12 Russia to close borders to curb coronavirus, *The Moscow Times*, available at www.the moscowtimes.com/2020/03/28/russia-to-close-borders-to-curb-coronavirus-a69785, accessed on May 15, 2020

13 Duclos M, Is COVID-19 a game-changer for Russia?, available at www.institutmont-aigne.org/en/blog/covid-19-game-changer-russia, accessed on May 13, 2020

14 Coronavirus: City restrictions and social support, available at www.sobyanin.ru/koronavirus-ogranichenie-peredvizheniya-i-sospodderzhka-grazhdan, accessed on May 15, 2020

15 Coronavirus disease 2019 (COVID-19) situation report – 86, WHO, available at www.who.int/docs/default-source/coronaviruse/situation-reports/20200415-sitrep-86-covid-19.pdf?sfvrsn=c615ea20_6, accessed on May 15, 2020

16 Coronavirus disease 2019 (COVID-19) situation report – 101, available at www.who.int/docs/default-source/coronaviruse/situation-reports/20200430-sitrep-101-covid-19.pdf?sfvrsn=2ba4e093_2, accessed on May 13, 2020

17 Coronavirus: Russian PM Mishustin tests positive for virus, *BBC*, available at www.bbc.com/news/world-europe-52491205, accessed on May 13, 2020

18 Coronavirus disease (COVID-19) situation report – 113, available at www.who.int/docs/default-source/coronaviruse/situation-reports/20200512-covid-19-sitrep-113.pdf?sfvrsn=feac3b6d_2, accessed on May 13, 2020

19 Coronavirus: Russia now has second highest virus case total, *BBC*, available at www.bbc.com/news/world-europe-52631045, accessed on May 15, 2020

20 Coronavirus: Putin eases Russian lockdown as cases rise, *BBC*, available at www.bbc.com/news/world-europe-52620015, accessed on May 15, 2020

21 Global report: Russia becomes Europe's coronavirus hotspot, *The Guardian*

22 Coronavirus disease (COVID-19) situation report – 116, WHO

23 Graphics: Why there was a sudden spike in Russia's coronavirus cases, available at https://news.cgtn.com/news/2020-04-27/Graphics-Why-there-was-a-sudden-spike-in-Russia-s-coronavirus-cases–Q1y5CXZ1gk/index.html, accessed on May 15, 2020

24 Coronavirus: Turkey sets strict measures as cases soar, *BBC*, available at www.bbc.com/news/world-europe-52185497, accessed on May 13, 2020

25 Turkey remains firm, calm as first coronavirus case confirmed, *Daily Sabah*, available at www.dailysabah.com/turkey/turkey-remains-firm-calm-as-first-coronavirus-case-confirmed/news, accessed on May 14, 2020

26 The number of cases rose to five in Turkey coronavirus, 9 countries more flights stopped, *Sputnik*, available at https://tr.sputniknews.com/turkiye/202003131041591545-koca-koronavirusun-turkiyedeki-seyri-diger-ulkelere-kiyasla-bir-avantaj-sunmakta/, accessed on May 14, 2020

27 Turkey's Coronavirus Scientific Advisory Board to convene today, available at https://ilkha.com/english/health-life/turkey-s-coronavirus-scientific-advisory-board-to-convene-today-7904, accessed on May 15, 2020

28 Turkey and Pakistan close borders with Iran over coronavirus deaths, *The Guardian*, available at www.theguardian.com/world/2020/feb/23/turkey-and-pakistan-close-borders-with-iran-over-coronavirus-deaths, accessed on May 15, 2020

29 REFILE-Turkey confirms first coronavirus death, more than doubles cases to 98, *Reuters*, available at https://in.reuters.com/article/health-coronavirus-turkey/turkey-confirms-first-coronavirus-death-more-than-doubles-cases-to-98-idINL8N2BA4O1, accessed on May 14, 2020

30 Coronavirus disease 2019 (COVID-19) situation report – 71, WHO

31 COVID-19 deaths in Turkey rise to 277, cases reach 15,679, *Daily Sabah*, available at www.dailysabah.com/turkey/covid-19-deaths-in-turkey-rise-to-277-cases-reach-15679/news, accessed on May 14, 2020
32 Ibid.
33 Stricter measures in Turkey as cases rise sharply, available at www.msn.com/en-us/news/world/stricter-measures-in-turkey-as-cases-rise-sharply/ar-BB12dH3I, accessed on May 15, 2020
34 Coronavirus: Turkey sets strict measures as cases soar, *BBC*
35 Coronavirus disease 2019 (COVID-19) situation report – 101, WHO
36 Coronavirus disease (COVID-19) situation report – 116, WHO
37 Ibid.
38 COVID-19 pandemic in Brazil, available at https://en.wikipedia.org/wiki/COVID-19_pandemic_in_Brazil, accessed on May 13, 2020
39 Picheta R, Darlington S, Hallam J, Brazilian president's press secretary tests positive for coronavirus, days after meeting Trump, *CNN*, available at https://edition.cnn.com/2020/03/12/americas/brazil-bolsonaro-coronavirus-aide-scli-intl/index.html, accessed on May 13, 2020
40 Coronavirus disease 2019 (COVID-19) situation report – 59, WHO, available at www.who.int/docs/default-source/coronaviruse/situation-reports/20200319-sitrep-59-covid-19.pdf?sfvrsn=c3dcdef9_2, accessed on May 13, 2020
41 Coronavirus: First Brazil death 'earlier than thought', *BBC*, available at www.bbc.com/news/world-latin-america-52638352, accessed on May 13, 2020
42 Ibid.
43 Coronavirus disease 2019 (COVID-19) situation report – 71, WHO
44 COVID-19 in Brazil: 'So what?' *Lancet*, available at www.thelancet.com/journals/lancet/article/PIIS0140-6736(20)31095-3/fulltext, accessed on May 13, 2020
45 First Yanomami COVID-19 death raises fears for Brazil's indigenous peoples, *The Guardian*, available at www.theguardian.com/world/2020/apr/10/first-yanomami-covid-19-death-brazl-indigenous, accessed on May 13, 2020
46 Ibid.
47 Coronavirus disease 2019 (COVID-19) situation report – 101, WHO
48 Coronavirus disease (COVID-19) situation report – 116, WHO
49 Kelly B, Amazon city resorts to mass graves as Brazil COVID-19 deaths soar, *Reuters*, available at https://in.reuters.com/article/health-coronavirus-brazil/amazon-city-resorts-to-mass-graves-as-brazil-covid-19-deaths-soar-idINKBN22C0D5, accessed on May 13, 2020
50 Ibid.
51 COVID-19 in Brazil: 'So what?' *Lancet*
52 Coronavirus: Brazil records highest daily rise in deaths, *BBC*, available at www.bbc.com/news/world-latin-america-52644339, accessed on May 13, 2020
53 Ibid.
54 Bolsonaro allows Brazilians to go to gyms, hair salons, as COVID-19 cases spike, *France 24*, available at www.france24.com/en/20200512-bolsonaro-allows-brazilians-to-go-to-gyms-and-hair-salons-as-covid-19-cases-spike, accessed on May 15, 2020
55 Brazil coronavirus cases hit daily record as Bolsonaro pressures, *Reuters*, available at https://in.reuters.com/article/health-coronavirus-brazil/brazil-coronavirus-cases-hit-daily-record-as-bolsonaro-pressures-ceos-idINKBN22R0K0, accessed on May 15, 2020
56 Bolsonaro fires popular health minister after dispute over coronavirus response, *The Guardian*, available at www.theguardian.com/world/2020/apr/16/bolsonaro-brazil-president-luiz-mandetta-health-minister, accessed on May 15, 2020
57 Brazil's star justice minister Sérgio Moro resigns in blow to Jair Bolsonaro, *The Guardian*, available at www.theguardian.com/world/2020/apr/24/brazil-justice-minister-sergio-moro-resigns-jair-bolsonaro, accessed on May 15, 2020
58 Coronavirus disease (COVID-19) situation report – 116, WHO
59 Ibid.

60 Wright R, How Iran became a new epicenter of the coronavirus outbreak, *New Yorker*, available at www.newyorker.com/news/our-columnists/how-iran-became-a-new-epicenter-of-the-coronavirus-outbreak, accessed on May 7, 2020

61 Ibid.

62 Ibid.

63 12 dead and up to 61 infected with coronavirus in Iran – Deputy health minister, *Reuters*, available at www.reuters.com/article/china-health-iran/12-dead-and-up-to-61-infected-with-coronavirus-in-iran-deputy-health-minister-idUSC6N2A401P, accessed on May 15, 2020

64 Coronavirus: Iran steps up efforts as 23 MPs said to be infected, *The Guardian*, available at www.theguardian.com/world/2020/mar/03/iran-steps-up-coronavirus-efforts-as-23-mps-said-to-be-infected, accessed on May 15, 2020

65 Coronavirus pandemic 'could kill millions' in Iran, *Al Jazeera*, available at www.aljazeera.com/news/2020/03/coronavirus-pandemic-kill-millions-iran-200317135500255.html, accessed on May 7, 2020

66 Iran's deputy health minister, seen sweating at briefing, has coronavirus, *Irish Times*, available at www.irishtimes.com/news/health/iran-s-deputy-health-minister-seen-sweating-at-briefing-has-coronavirus-1.4184607, accessed on May 15, 2020

67 Coronavirus disease 2019 (COVID-19) situation report – 41, WHO, available at www.who.int/docs/default-source/coronaviruse/situation-reports/20200301-sitrep-41-covid-19.pdf?sfvrsn=6768306d_2, accessed on May 15, 2020

68 Coronavirus disease 2019 (COVID-19) situation report – 55, WHO

69 Coronavirus disease 2019 (COVID-19) situation report – 53, WHO, available at www.who.int/docs/default-source/coronaviruse/situation-reports/20200313-sitrep-53-covid-19.pdf?sfvrsn=adb3f72_2, accessed on April 9, 2020

70 Coronavirus disease 2019 (COVID-19) situation report – 101, WHO

71 Iran leader refuses US help: Cites coronavirus conspiracy theory, *Al Jazeera*, available at www.aljazeera.com/news/2020/03/iran-leader-refuses-cites-coronavirus-conspiracy-theory-200322145122752.html, accessed on May 7, 2020

72 WHO to start coronavirus testing in rebel Syria; Iran raises efforts, official says, *Reuters*, available at www.reuters.com/article/us-health-coronavirus-mideast/who-to-start-coronavirus-testing-in-rebel-syria-iran-raises-efforts-official-says-idUSKBN2133PK, accessed on May 15, 2020

73 Tajdin B, Adamou L, Coronavirus: Are the bodies of victims undermining Iran's official figures? *BBC Persian*, available at www.bbc.com/news/world-middle-east-52223193, accessed on May 7, 2020

74 Ibid.

75 Coronavirus disease (COVID-19) situation report – 116, WHO

76 Ibid.

77 According to WHO classification of various regions of the world, Egypt and Morocco are considered as part of Eastern Mediterranean region even if they geographically lie in continent of Africa

78 Coronavirus disease 2019 (COVID-19) situation report – 37, WHO, available at www.who.int/docs/default-source/coronaviruse/situation-reports/20200226-sitrep-37-covid-19.pdf?sfvrsn=2146841e_2, accessed on May 15, 2020

79 A second COVID-19 case is confirmed in Africa, WHO, available at www.afro.who.int/news/second-covid-19-case-confirmed-africa, accessed on May 15, 2020

80 Ibid.

81 Coronavirus disease 2019 (COVID-19) situation report – 71, WHO

82 Coronavirus disease (COVID-19) situation report – 116, WHO

83 Ibid.

84 Ibid.

85 COVID-19 cases top 10 000 in Africa, WHO, available at www.afro.who.int/news/covid-19-cases-top-10-000-africa#:~:text=Reaching%20the%20continent%20through%20travellers,countries%20have%20reported%20cases, accessed on May 15, 2020

86 Ibid.

87 How the spread of coronavirus is testing Africa, *BBC*, available at www.bbc.com/news/world-africa-52230991, accessed on May 15, 2020

88 Ibid.

89 Africa facing a quarter of a billion coronavirus cases, WHO predicts, *The Guardian*, available at www.theguardian.com/global-development/2020/may/15/africa-facing-a-quarter-of-a-billion-coronavirus-cases-who-predicts, accessed on May 15, 2020

90 Ibid.

91 Coronavirus: WHO warns 190,000 could die in Africa in one year, *BBC*, available at www.bbc.com/news/world-africa-52587408, accessed on May 9, 2020

92 Coronavirus could 'smoulder' in Africa for several years, WHO warns, *The Guardian*, available at www.theguardian.com/world/2020/may/08/coronavirus-could-smoulder-in-africa-for-several-years-who-warns, accessed on May 15, 2020

93 Coronavirus disease (COVID-19) situation report – 116, WHO

94 Novel coronavirus (2019-nCoV) situation report – 10, WHO, available at www.who.int/docs/default-source/coronaviruse/situation-reports/20200130-sitrep-10-ncov.pdf?sfvrsn=d0b2e480_2, accessed on May 15, 2020

95 Rawat M, Coronavirus in India: Tracking country's first 50 COVID-19 cases; what numbers tell, *India Today*, available at www.indiatoday.in/india/story/coronavirus-in-india-tracking-country-s-first-50-covid-19-cases-what-numbers-tell-1654468-2020-03-12, accessed on May 7, 2020

96 Coronavirus disease (COVID-19) situation report – 116, WHO

97 COVID-19: India well-equipped, says Health Minister Harsh Vardhan, *The Hindu*, available at www.thehindu.com/news/national/india-well-equipped-harshvardhan/article31312701.ece, accessed on May 17, 2020

98 PM Narendra Modi announces 3-week national lockdown from March 24 midnight, *Deccan Herald*, available at www.deccanherald.com/national/pm-narendra-modi-announces-3-week-national-lockdown-from-march-24-midnight-817221.html, accessed on May 15, 2020

99 Coronavirus: India extends limited lockdown till May 17, tally climbs to 35,365, *India Today*, available at www.indiatoday.in/india/story/coronavirus-india-extends-limited-lockdown-till-may-17-tally-climbs-to-35-365-1673493-2020-05-02, accessed on May 15, 2020

100 Lockdown 4.0 to remain in force till May 31, MHA issues new guidelines for fourth phase, *India Today*, available at www.indiatoday.in/india/story/lockdown-4-0-india-covid-coronavirus-outbreak-guidelines-may31-rules-green-orange-red-containment-zones-updates-1678972-2020-05-17, accessed on May 17, 2020

101 Coronavirus: India defiant as millions struggle under lockdown, *BBC*, available at www.bbc.com/news/world-asia-india-52077395, accessed on May 7, 2020

102 India has tremendous capacity in eradicating coronavirus pandemic: WHO, *The Economic Times*, available at https://economictimes.indiatimes.com/news/politics-and-nation/india-has-tremendous-capacity-in-eradicating-coronavirus-pandemic-who/articleshow/74788341.cms, accessed on May 7, 2020

103 India scores high on COVID-19 response tracker made by Oxford University, *India Today*, available at www.indiatoday.in/india/story/india-scores-high-on-covid-19-response-tracker-made-by-oxford-university-1665573-2020-04-10, accessed on May 7, 2020

104 Lockdown 2.0: A complete list of activities permitted after April 20, *The Hindu Business Line*, available at www.thehindubusinessline.com/news/national/lockdown-20-a-complete-list-of-activities-permitted-after-april-20/article31375679.ece, accessed on May 15, 2020

105 Lockdown could be relaxed in some places after April 20: PM Modi, *Live Mint*, available at www.livemint.com/news/india/lockdown-could-be-relaxed-in-some-places-after-april-20-pm-modi-11586833018103.html, accessed on May 15, 2020

106 COVID-19 Karnataka wrap: Traffic jam in capital following lockdown relaxation; 37 new cases in state; two deaths, *The Indian Express*, available at https://indianexpress.com/article/cities/bangalore/covid-19-karnataka-wrap-traffic-jam-in-capitfollow

ing-lockdown-relaxation-37-new-cases-in-state-two-deaths-6393817/, accessed on May 15, 2020

107 Coronavirus disease (COVID-19) situation report – 116, WHO

108 Coronavirus: WHO praises Singapore's containment of COVID-19 outbreak, available at www.straitstimes.com/singapore/health/coronavirus-who-praises-singapores-containment-of-covid-19-outbreak, accessed on May 16, 2020

109 Coronavirus disease (COVID-19) situation report – 116, WHO

110 Public health responses to COVID-19 outbreaks on cruise ships – Worldwide, February–March 2020, CDC, *MMWR*, available at www.cdc.gov/mmwr/volumes/69/wr/mm6912e3.htm, accessed on April 10, 2020

111 Ibid.

112 Coronavirus disease 2019 (COVID-19) situation report – 32, WHO, available at www.who.int/docs/default-source/coronaviruse/situation-reports/20200221-sitrep-32-covid-19.pdf?sfvrsn=4802d089_2, accessed on May 16, 2020

113 www.businessinsider.in/science/news/how-the-failed-quarantine-of-the-diamond-princess-cruise-ship-started-with-10-coronavirus-cases-and-ended-with-more-than-630/articleshow/74267225.cms, accessed on April 10, 2020

114 Coronavirus: Passengers leave Diamond Princess amid criticism of Japan, *BBC*, available at www.bbc.com/news/world-asia-51555420, accessed on April 10, 2020

115 Public health responses to COVID-19 outbreaks on cruise ships – Worldwide, February–March 2020, CDC, *MMWR*

116 Ministry of Health, Labour and Welfare. *About New Coronavirus Infections* [Japanese], Tokyo, Japan: Ministry of Health, Labour and Welfare, 2020, available at www.mhlw.go.jp/stf/seisakunitsuite/bunya/0000164708_00001.htmlexternal icon, accessed on April 10, 2020

117 Trapped on a cruise ship by the coronavirus: when is the breakfast? *New York Times*, available at www.nytimes.com/2020/02/05/world/asia/japan-coronavirus-cruise-ship.html, accessed on May 16, 2020

118 Crew of virus-hit ship: 'We all are really scared', *CNN*, available at https://edition.cnn.com/asia/live-news/coronavirus-outbreak-02-13-20-intl-hnk/h_d05fee7e5eeda0b19d047bba537b0260, accessed on May 16, 2020

119 Public health responses to COVID-19 outbreaks on cruise ships – Worldwide, February–March 2020, CDC, *MMWR*

120 Ibid.

121 WHO's Nabarro: We must learn to live with COVID-19, *BBC*, available at www.bbc.com/news/av/health-52369969/who-s-nabarro-we-must-learn-to-live-with-covid-19, accessed on May 16, 2020

6

COVID-19

The signs of death

COVID-19 has posed critical challenges for everyone, especially for the frontline workers in public health, the medical community and other essential service workers. Dr Wang Li and his colleagues presented early data about COVID-19 which provided medical professionals much sought-after information about this deadly pandemic.[1] Since then, the pandemic has been progressing from one country to another, leaving behind death and misery.

In this fight against the deadly virus, however, these frontline soldiers are getting infected and are falling sick while doing their duties. So it becomes imperative that the war is fought by all of us together at the individual level as well. To do so, it is most important for us to know the details of the disease, its symptoms, how it is transmitted and what treatment is available for treating a patient infected with COVID-19.

Recognizing a case of COVID-19

An early detection of COVID-19 in a person is important for being able to offer timely treatment to the patient, which can go a long way towards saving the life of the person and also controlling the spread of the disease to others. The critical question therefore is, how do we recognize a case of COVID-19? In the early stages of the illness, the majority of patients infected with the novel coronavirus show symptoms similar to those of the common cold or a flu. Therefore, it becomes crucial to differentiate between COVID-19 and the common cold. The WHO has provided guidelines for case definition,[2] which can help in recognizing whether a patient has been infected by the novel coronavirus.

DOI: 10.4324/9781003345091-6

Suspect case

There are many people who have fever and cough, but that does not mean that all of them are infected with coronavirus. As per the WHO guidelines, a person is a suspect case if they exhibit any of the following features:[3]

1   A patient with acute respiratory illness (fever and at least one sign/symptom of respiratory disease (e.g. cough, shortness of breath), with no other aetiology[4] that fully explains the clinical presentation and a history of travel to or residence in a country/area or territory reporting local transmission of COVID-19 during the 14 days prior to symptom onset.
2   A patient with any acute respiratory illness and who has been in contact with a confirmed or probable COVID-19 case (see definition of contact) in the 14 days prior to onset of symptoms.
3   A patient with severe acute respiratory infection (fever and at least one sign/symptom of respiratory disease (e.g. cough, shortness breath) who requires hospitalization and who has no other aetiology that fully explains the clinical presentation.

Probable case

Similarly, a suspect case for whom testing for COVID-19 is inconclusive becomes a probable case – 'inconclusive' being the result of the test reported by the laboratory.

Confirmed case

A person with laboratory confirmation of COVID-19 infection, irrespective of clinical signs and symptoms, is a confirmed case of COVID-19.

Contact

During this pandemic, the term 'contact' has been widely used in the media. The virus is moving from person to person in the community. So anybody may be exposed to a person who has COVID-19. The problem becomes more acute because many of the infected people are asymptomatic. A healthy person becomes the contact if they are exposed to a patient who has COVID-19. The WHO defines contact as a person who has experienced any one of the following exposures during the two days before and the 14 days after the onset of symptoms of a probable or confirmed case:

1   Face-to-face contact with a probable or confirmed case within 2 metres and for more than 15 minutes

2　Direct physical contact with a probable or confirmed case

3　Direct care for a patient with probable or confirmed COVID-19 case without using proper PPE

4　Other situations as indicated by local risk assessments.

Basic clinical aspects

The first and the most important step in identifying a case of COVID-19 is the history of travel. If a person who has travelled to countries like China, the US, Italy, France, Spain and the UK develops cough and fever subsequently, they most likely have COVID-19. Hence, the most important next step is to keep a safe distance from such a person and to inform the healthcare authorities for a proper diagnosis and treatment. History of travel was extremely important in the early months of the pandemic. However, since the pandemic has spread to every part of the world, travel history has lost some of its relevance. It is still crucial in those countries that have not yet been badly affected.

Symptoms

The COVID-19 patient may be asymptomatic, which means they do not even feel that they are sick. At the other end, there are those who develop respiratory failure and require ventilator support in ICU, sepsis, septic shock and multiple organ dysfunction syndrome (MODS).[5]

The most common symptoms[6,7,8] are fever, dry cough, shortness of breath, body ache and fatigue. Fever may be associated with chills. Less-common symptoms are anorexia, sputum production, sore throat, confusion, dizziness, headache, runny nose, chest pain, presence of blood in cough, skin rash or discolouration of fingers and toes, new loss of taste or smell, diarrhoea, nausea/vomiting and abdominal pain. Certain neurological features like tingling or numbness in the hands and feet, dizziness, confusion, delirium, seizures and stroke are seen in some patients.[9] Gastrointestinal symptoms, such as loss of appetite, nausea, vomiting, diarrhoea and abdominal pain or discomfort might start before other symptoms, such as fever, body ache and cough. Heart involvement, leading to myocarditis,[10] myocardial infarction[11] or arrhythmias,[12] have been reported in COVID-19 patients.[13] Eye symptoms like itching, redness, tearing, discharge and foreign body sensation are also found in COVID-19 patients. Conjunctivitis has been reported in some patients.[14]

For the majority of patients, COVID-19 is a nuisance like the common cold, but unfortunately for some patients, it quickly turns into a struggle between life and death. Some patients experience severe cough and describe it as, for example, 'I was coughing like I was going to die'.[15] Others lost their voice due to incessant cough. At bedtime for some, there was coughing, and upon waking up, breathing became difficult for many others. A 55-year-old resident of Washington complained of heaviness of the chest as if an elephant were standing on his chest.[16]

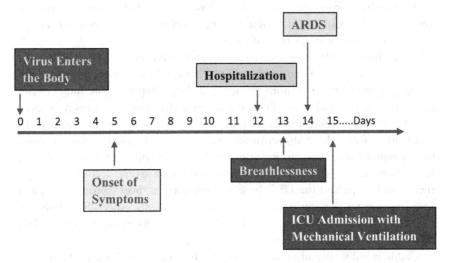

FIGURE 6.1 A Timeline for the Development of Various Symptoms of COVID-19

Source: Author.

Some others described their condition as taking every breath was a battle. Patients complain of high-grade fever and constant dry cough. Some of them require hospitalization and after a few days develop breathlessness, which further leads to the development of ARDS.[17] The patient requires ICU care and mechanical ventilation. The timeline for the development of various features is shown in Figure 6.1.

P. H., a 48-year-old university professor at Busan, South Korea, was extremely health conscious. On February 21, 2020, P. H. said that he felt a 'very mild sore throat and very mild dry coughing'. On February 24, he had breathing problems early in the morning. His fear came true. He tested positive for coronavirus. He was later admitted to a negative-pressure room in the quarantine section of Kosin University Gospel Hospital's ICU. He received oxygen and a CT scan of his chest was done. He had developed severe chest pain: 'I felt a burning pain in my chest and stomach, although I was not sure if it was because of the drugs I took or the virus, . . . I had a slight fever and my condition fluctuated. At first I felt as if some heavy iron plate . . . was pressing on my chest. The stabbing pain gradually eased to the point it felt as if someone was squeezing my chest hard'. He recovered from the ordeal and later said, 'I was naive and stupid to think that [the outbreak] is not my problem. Yes, as usual, I was stupidly overconfident'.[18]

Shortness of breath is one of the early signs of danger. Often these patients require oxygen and deteriorate rapidly. C., 28, from South Africa, wrote, 'The best way to describe it is, when you are at a high altitude, you struggle to breathe. . . . Within days, it fluctuates. You get chilled and later on you feel better. . . . The worst for me was last week. . . . I was really short of breath. . . . I called my doctor'.[19] An Italian COVID-19 survivor said, 'You are there on the bed attached to the oxygen, you feel breathless and you understand it may be your last breath'.[20]

'It's like being run over by a rammer. Everything aches. Your muscles ache. Your joints ache', said Federico Gutiérrez, the chief of cardiology at La Paz Hospital, who survived the attack of coronavirus.[21]

Approximately 80% of patients present with mild illness, 14% present with severe illness and 5% present with critical illness.[22] Severe illness is seen mainly in the presence of other comorbidities, like diabetes, hypertension, lung diseases, kidney dysfunction and so on. The most common diagnosis in patients with severe COVID-19 is severe pneumonia.[23] Kidney involvement is frequent in COVID-19; under 40% of cases have abnormal proteinuria[24] upon hospital admission. Acute kidney injury is common among critically ill patients with COVID-19, affecting approximately 20%–40% of patients admitted to intensive care, according to experiences in Europe and the US.[25] In severely ill patients, liver damage with raised liver enzymes are also reported.[26] However, many patients, especially in India, are found to be absolutely asymptomatic and are thought to have extensively spread the disease in the community (Figure 6.2).

Children and young adults are affected by the virus, though their rate of infection and severity of disease are much lower than those of older people. However, this should not lead to a false sense of security. Children are typically asymptomatic or present with mild symptoms like fever, cough, sore throat and so on.[27, 28] There have been several cases of Kawasaki disease, an overlap syndrome of toxic shock

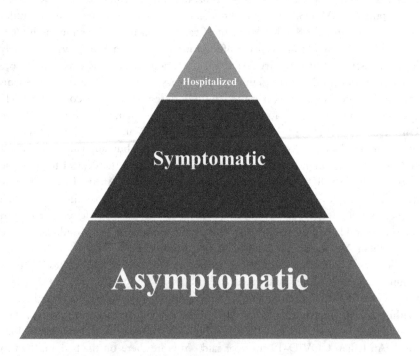

FIGURE 6.2 Clinical Presentation of COVID-19

and Kawasaki syndrome[29] in children with COVID-19. Presentations include fever, conjunctivitis, a polymorphous and blanching rash, tongue involvement and swelling of hands and feet.[30] Neonates and infants have also been reported to have developed COVID-19 and to have experienced diarrhoea.[31] Pregnant women with COVID-19 have similar features to those reported for non-pregnant adults.[32,33]

However, COVID-19 takes a toll on the mental health of patient as well. Pain, loneliness and panic affect most of patients. For example, 65-year-old cardiologist F. B. had COVID-19 and spent eight days 'isolated from the world' at Rome's Policlinico Umberto I hospital. The most difficult thing for him were the nights alone with his fears: 'I couldn't sleep, anxiety invaded the room . . . nightmares came, death prowled. . . . I was afraid of dying without being able to cling on to the hands of my family and friends, despair overcame me'.[34]

Everybody is worried and keeping away from others to avoid getting infected. A huge number of asymptomatic cases have made it impossible for anybody to know whether the person standing next to them has COVID-19 or not. A simple bout of fever or a few bouts of cough are enough to worry people. The accompanying flowchart (Figure 6.3) is a guide for action after coming into contact with a person who may have COVID-19.

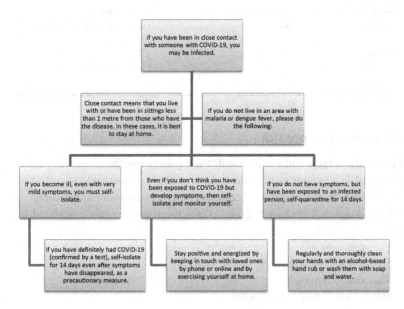

FIGURE 6.3   What to Do If You Come into Close Contact with a COVID-19-Infected Person

Comparison between COVID-19 and the common cold

Initially, the common cold and COVID-19 have similar features, and it may become a bit difficult to differentiate between the two. People all over the world are worried about coughing and having a mild fever. Suggestions from the European Centre for Disease Prevention and Control help to distinguish between them (Table 6.1).[35]

As reported in the *Guardian*, A., 41, developed COVID-19 and was admitted to a hospital in Edinburgh. She developed a fever, a constant cough, abdominal and chest pain, shortness of breath and the loss of her sense of smell and taste. She had coughed up blood as well and developed tightness in her chest. She improved subsequently and was discharged.[36]

TABLE 6.1 Comparison between COVID-19 and Common Cold

Features	Common Cold	COVID-19
Fever	Sometimes, may be absent	Yes, may be high, not responding to paracetamol
Cough	Sometimes, maybe dry or with phlegm	Persistent dry cough
Sneezing	Common	Uncommon
Headache	Sometimes	Common and often severe
Runny nose	Common	Uncommon
Loss of smell	Sometimes	Sometimes
Body ache	Sometimes	Common and sometimes severe
Fatigue	Sometimes	Common and severe
Response to cold medicine	Good	Does not respond

What to do if you are sick

If a person has a fever and a cough with chest pain or breathlessness, they may have COVID-19. The majority of patients have mild symptoms and recover at home. The most important first step is to contact a healthcare provider immediately. Study the accompanying flowchart (Figure 6.4) for the following steps suggested by the CDC:[37]

- Stay at home except to get medical care.
- Take care of yourself.
- Stay in touch with your doctor.
- Avoid public transport.
- Stay away from others, in a separate room with an attached toilet/bathroom.
- Monitor your symptoms, and inform your doctor.
- Wear a mask when you are near other people or animals, including pets.
- Cover your mouth and nose while coughing or sneezing.
- Clean hands frequently with soap and water.
- Avoid sharing personal items.
- Clean all high-touch surfaces every day.

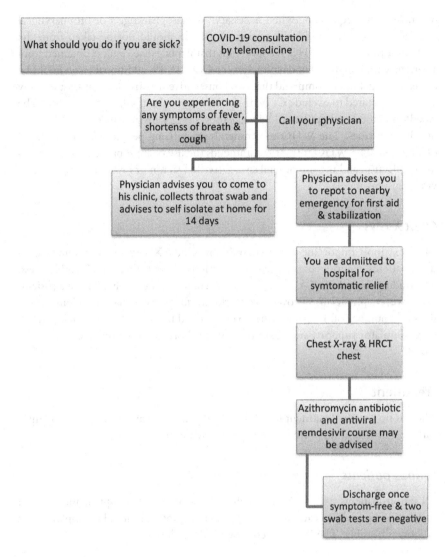

FIGURE 6.4   What Should You Do If You Fall Sick?

Laboratory tests

Molecular testing is required to confirm the diagnosis. There is a shortage of diagnostic kits in almost all the countries affected by the COVID-19 pandemic. Hence, diagnostic tests should be performed according to guidance issued by local health authorities and should adhere to appropriate biosafety practices.

The nucleic acid amplification test

A nucleic acid amplification test, such as reverse transcriptase polymerase chain reaction (RT-PCR), is used to diagnose SARS-CoV-2 infection, with confirmation by nucleic acid sequencing when necessary.[38] A nasopharyngeal or oropharyngeal swab is collected for molecular testing. A sample should not be taken from the nostrils or tonsils. The WHO recommends that if a negative result is obtained from a patient with a high index of suspicion for COVID-19, additional tests should be done. Its guidelines recommend that two consecutive negative tests (at least one day apart) are required to exclude COVID-19. Decisions about whom to test should be based on clinical and epidemiological factors, along with guidance issued by local health authorities. The WHO recommends prioritizing people with a likelihood of infection. The WHO further recommends the use of appropriate PPE for specimen collection and the use of viral swabs (sterile Dacron or rayon, not cotton) and viral transport media.

Chest X-rays

The next most important test is chest X-ray. Chest X-rays reveal pneumonia in those patients experiencing severe cough with high-grade fever and breathlessness. The findings in a chest X-ray enable a physician to appropriately manage a patient.

However, physicians all over the world are facing two main problems. First, the sheer number of the patients has overwhelmed hospitals. Second, sick patients need ventilator support, and there is an acute shortage of ventilators all over the world.

Treatment

There is no specific treatment for COVID-19 yet. The mainstay of management is early diagnosis and supportive care to relieve symptoms.

Serious patients

All serious patients with COVID-19 should be admitted to hospital, and some of them will require ICU care. Infection prevention protocol should be implemented as soon as possible. The WHO recommends the following:[39]

- Patients must be kept in well-ventilated rooms.
- Each patient should be kept in their own room.
- If a single room is not available, then keep patients in a single ward 2 metres apart.
- Use protective gear and single-use or disposable equipment.
- Limit the number of healthcare workers (HCWs).
- Consider keeping patients in a negative-pressure room.

All of these patients may require oxygen therapy. Fluid management is essential to prevent shock. Preventing and managing complications like ARDS, sepsis and septic shock should be given utmost priority.

Starting antibiotics to prevent bacterial infections and the choice of antibiotics will depend upon local protocol. Antipyretics and analgesics are used to treat fever and pain. Some authorities have advised against the use of ibuprofen.[40] The NHS in the UK recommends paracetamol as the drug of choice.[41]

Older patients, especially with comorbidities, require a lot of care because they are most likely to deteriorate rapidly. At the same time, young patients who appear well also may deteriorate rapidly. In some patients, it has been observed that they collapse suddenly and require mechanical ventilation.

Corticosteroids have been used in some patients with COVID-19; however, they have been found to be ineffective and are not recommended. Similarly, many experimental therapies have been tried on compassionate grounds, like interferon, chloroquine/hydroxychloroquine, azithromycin, lopinavir/ritonavir and convalescent plasma, with doubtful results. On May 1, 2020, the US Food and Drug Administration issued an emergency use authorization for the investigational antiviral drug remdesivir for the treatment of suspected or laboratory-confirmed COVID-19 in adults and children hospitalized with severe disease.[42]

Less-serious patients

Ideally, patients with mild illness who have risk factors for poor outcomes (i.e. over 60, presence of comorbidities) should also be prioritized for hospital admission.[43] However, due to the shortage of beds, it is recommended by the WHO[44] that confirmed cases without any comorbidities and risk factors be treated at home. For home care, there should be some family member to look after the patient. A healthcare worker must visit this patient regularly and keep in touch with the local health authority. Patients and their family members must strictly follow the infection prevention protocol. If symptoms worsen, they should be immediately shifted to a well-equipped hospital. Two negative test results (on samples collected at least 24 hours apart) are required before the patient can be released from home isolation.

R., a third-year engineering student at Imperial College London complaining of coughing red phlegm, exhaustion, night sweats and nausea, was asked to stay at home in the UK and was asked to report breathlessness if that symptom developed. He said, 'I couldn't get out of bed when the fever was at its worst. I lost my appetite and could barely drink water'.[45]

Infection prevention

Infection prevention protocol, recommended by the WHO,[46] should be implemented for suspected or confirmed cases of COVID-19. Every healthcare worker (HCW) must follow hand and respiratory hygiene, use appropriate PPE and practise safe waste management and sterilization of patient-care equipment. PPE effectiveness depends largely on adequate and regular supplies, adequate staff training, appropriate hand hygiene and appropriate human behaviour.[47,48]

Several respiratory hygiene measures should be followed:

- HCWs should ensure that all patients cover their mouths and noses and cough and sneeze into their flexed elbows.
- HCWs should offer a medical mask to suspected COVID-19 patients.

Every patient coming to hospital or a clinic must be advised to wash their hands with soap and running water before entering the premises. Every patient waiting outdoors for treatment should be asked to keep a minimum distance of 2 metres between them. Close contact must be avoided. HCWs must practise proper hand hygiene before and after examining every patient. Hand hygiene means cleaning hands with either alcohol-based hand rub or soap and water. Also, hospital and environmental surfaces should be regularly cleaned with water and detergent, and sodium hypochlorite should be applied for disinfection.

Doctors and nurses don a hazmat suit while taking care of patients. However, PPE kits are in short supply all over the world. Doctors and nurses are demanding PPE kits everywhere, and many are forced to look after patients without them. They feel anxiety and extreme panic as they are defenceless against COVID-19. In Spain, for example, at least one in seven of those infected is a healthcare worker, as many of them lack proper protective gear. But some have now recovered and are willing to go back to the frontline.[49]

Management of contacts

Every contact should self-isolate at home. They should stay in a separate room with an attached toilet/bathroom. Their clothes and utensils should be washed and cleaned separately. Every contact should be asked to regularly monitor their health. They should be asked to check their temperature at least twice per day. If they develop fever and/or cough, they must visit a designated treatment centre for testing COVID-19. While travelling, they must wear a face mask and keep a distance of at least 2 metres from everybody.

It has been reported that some patients test positive sometime after discharge (disease reactivation)[50] and hence will be a source of infection to others in the community. Data are not yet available regarding long-term complications among survivors of COVID-19. A vaccine is not yet available to prevent infection; however, efforts are being made in several countries to develop a safe and effective vaccine. Data are not yet available regarding the development and persistence of long-term immunity after recovery from COVID-19. The old adage 'prevention is better than cure' is applicable for COVID-19.

Besides health, COVID-19 has affected lives in various other ways. People's mental health has deteriorated. Stress and anxiety are taking their toll. People's social lives have been disrupted. People are not able to move out of their home due to lockdown. The world is facing a huge economic crisis. It is crucial to understand these issues and to look into measures to tackle them as soon as possible.

Notes

1 Li Q, Guan X, Wu P, et al., Early transmission dynamics in Wuhan, China, of novel coronavirus – Infected pneumonia, *N Engl J Med*, doi:10.1056/NEJMoa2001316

2 Coronavirus disease 2019 (COVID-19) situation report – 48, WHO, available at www.who.int/docs/default-source/coronaviruse/situation-reports/20200308-sitrep-48-covid-19.pdf?sfvrsn=16f7ccef_4, accessed on March 28, 2020

3 Ibid.

4 Aetiology means the cause, set of causes or manner of causation of a disease

5 Cascella M, Rajnik M, Cuomo A, et al., *Features, Evaluation and Treatment Coronavirus (COVID-19)*, available at www.ncbi.nlm.nih.gov/books/NBK554776/, accessed on March 28, 2020

6 Huang C, Wang Y, Li X, et al., Clinical features of patients infected with 2019 novel coronavirus in Wuhan, China, *Lancet*, February 15, 2020; 395(10223): 497–506

7 Wang D, Hu B, Hu C, et al., Clinical characteristics of 138 hospitalized patients with 2019 novel coronavirus-infected pneumonia in Wuhan, China, *JAMA*, February 7, 2020 [Epub ahead of print]

8 Li LQ, Huang T, Wang YQ, et al., 2019 novel coronavirus patients' clinical characteristics, discharge rate and fatality rate of meta-analysis, *J Med Virol*, March 12, 2020 [Epub ahead of print]

9 COVID-19 basics, available at www.health.harvard.edu/diseases-and-conditions/covid-19-basics, accessed on May 17, 2020

10 Inflammation of heart muscle

11 Heart attack due to blockage of coronary arteries

12 Irregular heart rhythm

13 COVID-19 and cardiology, ESC, available at www.escardio.org/Education/COVID-19-and-Cardiology, accessed on May 18, 2020

14 Venkateswaran N, What are the ocular symptoms of COVID-19? Available at www.aao.org/editors-choice/what-are-ocular-manifestations-of-covid-19, accessed on May 18, 2020

15 'Coughing like I was going to die'. Here's what it's like to survive coronavirus in Wuhan, *Time*, available at https://time.com/5783838/coronavirus-symptoms-wuhan-survivor/, accessed on May 16, 2020

16 What does coronavirus feel like? Here's what survivors and patients say about it, available at www.sacbee.com/news/coronavirus/article241363476.html, accessed on May 16, 2020

17 Acute respiratory distress syndrome (ARDS) occurs when fluid builds up in the tiny, elastic air sacs (alveoli) in your lungs. People then experience breathlessness.

18 'I was stupidly overconfident': A South Korean coronavirus survivor's tale, *South China Morning Post*, available at www.scmp.com/week-asia/health-environment/article/3075170/i-was-stupidly-overconfident-south-korean-coronavirus, accessed on April 10, 2020

19 Pain, solitude, fear: Stories of surviving COVID-19, *The Times of India*, available at https://timesofindia.indiatimes.com/world/europe/pain-solitude-fear-stories-of-surviving-covid-19/articleshow/75025118.cms, accessed on May 16, 2020

20 www.euronews.com/2020/03/31/covid-19-survivors-bring-hope-and-purpose-to-spanish-hospital-workers, accessed on April 10, 2020

21 Ibid.

22 Novel Coronavirus Pneumonia Emergency Response Epidemiology Team, The epidemiological characteristics of an outbreak of 2019 novel coronavirus diseases (COVID-19) in China [in Chinese], *Zhonghua Liu Xing Bing Xue Za Zhi*, February 17, 2020; 41(2): 145–151

23 Clinical management of severe acute respiratory infection (SARI) when COVID-19 disease is suspected, WHO, March 2020

24 Loss of protein in urine

25 Ronco C, Reis T, Husain-Syed F, Management of acute kidney injury in patients with COVID-19, *Lancet*, available at www.thelancet.com/journals/lanres/article/PIIS2213-2600(20)30229-0/fulltext, accessed on May 18, 2020

26 Liver injury in COVID-19: Management and challenges, *Lancet*, available at www. thelancet.com/journals/langas/article/PIIS2468-1253(20)30057-1/fulltext, accessed on May 18, 2020

27 Shen KL, Yang YH, Diagnosis and treatment of 2019 novel coronavirus infection in children: A pressing issue, *World J Pediatr*, 2020 Jun; 16(3): 219–221, February 5, 2020

28 Wang XF, Yuan J, Zheng YJ, et al., Clinical and epidemiological characteristics of 34 children with 2019 novel coronavirus infection in Shenzhen [in Chinese], *ZhonghuaErKe Za Zhi*, February 17, 2020; 58(0): E008

29 Children experience fever, rash, swelling of the hands and feet, irritation and redness of the whites of the eyes, swollen lymph glands in the neck and irritation and inflammation of the mouth, lips and throat

30 Jones VG, Mills M, Suarez D, et al., COVID-19 and Kawasaki disease: Novel virus and novel case, *Hosp Pediatr*, 2020 Jun; 10(6): 537–540, 2020, doi:10.1542/hpeds.2020-0123

31 Liu W, Zhang Q, Chen J, et al., Detection of COVID-19 in children in early January 2020 in Wuhan, China, *N Engl J Med*, March 12, 2020; 382: 1370–1371, [Epub ahead of print]

32 Zhu H, Wang L, Fang C, et al., Clinical analysis of 10 neonates born to mothers with 2019-nCoV pneumonia, *TranslPediatr*, February 2020; 9(1): 51–60

33 Chen H, Guo J, Wang C, et al., Clinical characteristics and intrauterine vertical transmission potential of COVID-19 infection in nine pregnant women: A retrospective review of medical records, *Lancet*, March 7, 2020; 395(10226): 809–815

34 Pain, solitude, fear: Stories of surviving COVID-19, *The Times of India*

35 www.ecdc.europa.eu/en/covid-19/questions-answers, accessed on April 4, 2020

36 COVID-19 recoveries: 'It was the most terrifying experience of my life', *The Guardian*, available at www.theguardian.com/society/2020/apr/01/covid-19-recoveries-it-was-the-most-terrifying-experience-of-my-life, accessed on April 10, 2020

37 What to do if you are sick, CDC, available at www.cdc.gov/coronavirus/2019-ncov/if-you-are-sick/steps-when-sick.html, accessed on May 9, 2020

38 Laboratory testing for coronavirus disease 2019 (COVID-19) in suspected human cases, WHO, March 2020, available at https://apps.who.int/iris/handle/10665/331329, accessed on March 28, 2020

39 Infection prevention and control during health care when novel coronavirus (nCoV) infection is suspected, WHO, available at www.who.int/publications-detail/infection-prevention-and-control-during-health-care-when-novel-coronavirus-(ncov)-infection-is-suspected-20200125, accessed on May 16, 2020

40 Day M, COVID-19: Ibuprofen should not be used for managing symptoms, say doctors and scientists, *BMJ*, March 17, 2020; 368: m1086

41 Coronavirus disease 2019 (COVID 19), *BMJ Best Practice*, available at https://bestprac tice.bmj.com/topics/en-gb/3000168/diagnosis-approach, accessed on March 29, 2020

42 Coronavirus (COVID-19) update: FDA issues emergency use authorization for potential COVID-19 treatment, available at www.fda.gov/news-events/press-announcements/coronavirus-covid-19-update-fda-issues-emergency-use-authorization-potential-covid-19-treatment, accessed on May 9, 2020

43 Home care for patients with COVID-19 presenting with mild symptoms and management of their contacts, available at www.who.int/publications-detail/home-care-for-patients-with-suspected-novel-coronavirus-(ncov)-infection-presenting-with-mild-symptoms-and-management-of-contacts, accessed on March 29, 2020

44 Ibid.

45 COVID-19 recoveries: 'It was the most terrifying experience of my life', *The Guardian*

46 Infection prevention and control during health care when novel coronavirus (nCoV) infection is suspected, WHO

47 Infection prevention and control of epidemic- and pandemic-prone acute respiratory diseases in health care, WHO, March 29, 2020

48 How to put on and take off Personal Protective Equipment (PPE), Geneva: World Health Organization, 2008, available at www.who.int/csr/resources/publications/putontakeoffPPE/en/, accessed on March 29, 2020

49 www.euronews.com/2020/03/31/covid-19-survivors-bring-hope-and-purpose-to-spanish-hospital-workers, accessed on April 10, 2020

50 Chen D, Xu W, Lei Z, et al., Recurrence of positive SARS-CoV-2 RNA in COVID-19: A case report, *Int J Infect Dis*, April 2020; 93: 297–299, March 5, 2020; Xing Y, Mo P, Xiao Y, et al., Post-discharge surveillance and positive virus detection in two medical staff recovered from coronavirus disease 2019 (COVID-19), China, January to February 2020, *Euro Surveill*, March 2020; 25(10); Ye G, Pan Z, Pan Y, et al., Clinical characteristics of severe acute respiratory syndrome coronavirus 2 reactivation, *J Infect*, March 11, 2020 [Epub ahead of print]

7

IMPACT OF COVID-19

On people and the economy

The world is reeling from the impact of the COVID-19 pandemic, which is spreading across the globe, destroying the lives and livelihoods of thousands of people. Death, lockdown, loss of jobs, overwhelmed healthcare systems or lack of it, failure of governments in safeguarding the lives of their citizens, isolation and quarantine have resulted in extreme anxiety and fear.

COVID-19 has changed the way of life for millions of people across the globe. Social distancing, self-isolation, quarantine, curfew, travel bans and closures (of pubs, cinemas and malls) have completely disrupted social life. The cancellation and postponement of many important events, including that of the Tokyo Olympics, have left the world wondering what's next.

Economic devastation is equally widespread. City lockdowns, curfews and the closure of shops and industries have resulted in widespread job losses. The *Guardian* has reported that a record 36 million Americans have filed for unemployment in the past two months.[1] Some 20 million people lost their jobs in April, and the unemployment rate shot up to 14.7% from just 4.4% in March. People without a college education, African Americans and Latinos were the most affected. At the same time, migrant workers in India lost jobs in the cities and were forced to walk hundreds of miles to their homes in remote villages amid the lockdown. Markets all over the world have crashed, wiping out millions of dollars of investors' wealth.

Mental health

The pandemic has taken a toll on the mental health of almost everybody. Hardly anyone has remained untouched. The pandemic has prevented us from doing what we love to do, going where we want to be and being with whom we want to be with.[2] No one is exactly sure what is going to happen next or how bad the situation will ultimately turn out to be. People are worried about their own life and the lives

DOI: 10.4324/9781003345091-7

of their loved ones. For many, this is not easy to handle. Loss of sleep, nightmares, difficulty concentrating, family neglect, substance abuse and worsening existing health conditions are common all over. Overwhelming dread and panic have set in. According to the *Guardian*, Thomas Schaefer, the finance minister of Germany's Hesse state, apparently committed suicide after becoming 'deeply worried' over how to cope with the economic fallout from the coronavirus.[3] Similarly, a suspected coronavirus patient jumped from the seventh floor of Safdarjung Hospital, New Delhi, committed suicide.[4]

Globally, people who have anxiety and depression are on the rise. While those who can avail themselves of help can be treated by medicine, others at times do not recognize the problem or cannot consult a physician. The stigma attached to psychiatric disorders prevents many from getting proper treatment. The COVID-19 pandemic has caused both these groups to struggle immensely. It is natural to worry, but if it overwhelms your daily activities, including your thoughts and actions, then it is a matter for concern.

China experienced the SARS epidemic during 2002–2003, which killed many, led to economic crisis and had a deep impact on people's social lives and the mental health. The high death rate had a deep impact on the minds of Chinese people and many, including healthcare workers, developed depression and anxiety.[5] Many survivors developed post-traumatic stress disorder.[6] A similar challenge has been posed by the COVID-19 pandemic, not only to the people of China but to almost everybody else all over the world.

People react differently to stress. Reactions often depend on a variety of factors, which include socioeconomic status, accessibility to healthcare, education background and social and family support, among many others. The following people are the most vulnerable to stress:

- Children and teenagers
- Old people with a chronic disease
- Healthcare workers who are at the frontline in treating COVID-19 patients
- People who already experience psychological dysfunction
- The homeless and the poor
- Migrant workers
- People with drug addictions

The present mental health landscape has posed a great challenge to psychiatrists and mental health care workers all over the world. While a visit to a psychiatrist may address some issues, this may not work for everybody, because people react differently to stress and anxiety. It is more important now than ever for people to try to address their own stress and anxiety, seeking help whenever necessary.

It is important to have the right information and to stay away from fake news circulating on social media. Misinformation and sensationalistic coverage invoke a sense of doom, which could be quite harmful. It would help to trust information only from reputable and authentic sources like the WHO and the CDC and avoid relying heavily on social media, unverified news sources, message forwards and even avoid discussing the pandemic with people for too long.

Self-care becomes more important than just a keyword to make it through the pandemic. It is important not only to be mindful of health and hygiene to avoid the infection, but also to maintain mental hygiene. Small things like eating healthy, sleeping well, reading books, exercising and listening to music should be valued to reduce anxiety and distress: staying connected to people you love, reaching out to people you have been planning to make some time for and spending quality time with your family. Helping others in the community goes a long way to making a difference.

Looking after children

Children also develop anxiety, mainly by observing their parents. So be calm, and look after children. A calm demeanour will go a long way towards helping children. Look for the following signs of stress in children:[7]

- Excessive crying or irritation
- Lack of attention and concentration
- Poor scholastic performance
- Bedwetting
- Excessive worry or sadness
- Use of alcohol
- Insomnia
- Unhealthy eating
- Unexplained headaches

De-stress doctor

Healthcare workers are under severe physical and mental stress. Frequent night shifts, staying away from family, infrequent meals, failure to prevent death and worries about catching infection and transmitting it to loved ones are taking a great toll on their mental health. They need to relax and de-stress, as a single mistake on their part can cause irreparable damage to those they look after.

First of all, they must try to acknowledge that they are experiencing stress. Overlooking it is not going to help them. They should talk to their friends, colleagues, and family members to reduce anxiety. Experienced senior colleagues are often the best people to help reduce work-related stress. Allow some time for personal care and for family members. Listen to some music and do some light reading, if time permits. Keep social media contact to a minimum, or avoid it if possible. It is important to heal one's mind before healing others.

Social disruption

The coronavirus pandemic has affected everyone's life in some way or another. A significant proportion of population is either under lockdown, quarantine or home isolation or has been hospitalized with COVID-19. In several countries, offices are closed, schools and colleges are closed, industries have shut down and borders have been sealed. These unprecedented measures have resulted in wide-spread disruption to life and liberty and changed communities' social fabrics, and job loss has pushed millions into hunger and abject poverty.

COVID-19 coincided with the Chinese New Year holiday, which is one of the most celebrated times of the year in China.[8] During this festival, people go back to their hometown. To isolate people and stop the spread of the pandemic, the government initially extended the holiday. At the same time, Hong Kong Disneyland and Shanghai Disneyland were closed,[9] as were many other public places, to avoid crowds of people. This was the first disruption of social life during this pandemic, which was followed by numerous such happenings worldwide.

The closure of important tourist spots like the Taj Mahal, the Eiffel Tower, the Colosseum, the statue of Christ the Redeemer, the Statue of Liberty and Ellis Island and the sealing of borders and travel bans have disrupted the travel and tourism industry. Besides this, all over the world, numerous restaurants, cinema halls, shopping malls, pubs and markets have been closed to prevent public gathering. The closure of schools and colleges in various countries has disrupted academic schedules. Examinations and graduate-level entry examinations have been postponed, with students waiting to see how and when they will be rescheduled and conducted.

Event cancellations

The fear of coronavirus has resulted in the postponement or cancellation of many important major events all over the world. The London Book Fair[10] and the Geneva Motor Show 2020[11] were cancelled. Similarly, the American College of Cardiology cancelled the ACC's Scientific Session and the World Congress of Cardiology, scheduled to take place on March 28–30, 2020, in Chicago. Social media giant Facebook also cancelled its developer conference in San Jose because of coronavirus concerns. These unprecedented disruptions in peacetime have disrupted the plans of numerous people across the world.

Sporting event cancellations

To safeguard the life of spectators and athletes, many major sporting events have been either cancelled or postponed. The major casualty is the Olympic Games in Tokyo in 2020. Given the present circumstances and on the basis of the information provided by the WHO, the IOC president and the prime minister of Japan have concluded that the Games of the XXXII Olympiad in Tokyo must be rescheduled to a date in summer 2021, to safeguard the health of the athletes, everybody involved in the Olympic Games and the international community.[12] Now it has been rescheduled to July 23 to August 8, 2021.

Euro 2020 has been postponed and will now take place from June 11 to July 11, following the global coronavirus pandemic.[13] Similarly, all professional football in England, including the Premier League, was halted due to the coronavirus pandemic, in an effort to delay the spread of COVID-19.[14] Fear and concern for the safety of players, spectators and organizers led to the suspension of all five of the top leagues in Europe. Bundesliga and Ligue 1 join La Liga, Serie A and the Premier League in halting play for the time being. The UEFA Champions League and Europa League have also been suspended.[15] The majority of international cricket tournaments have been either cancelled or postponed due to COVID-19. The IPL has been postponed in India, and uncertainty looms over the forthcoming T20 World Cup.[16]

The coronavirus pandemic caused the cancellation of the entire grass court season and most of the clay court tennis campaign of 2020. The Wimbledon championships were cancelled by the AELTC on Wednesday for the first time since World War II. The 134th Championships will instead be staged from June 28 to July 11, 2021. The French Open, originally due to be played from May 24 to June 7, 2020, was rescheduled by the French tennis federation for September 2 to October 4, 2020. Many top tennis players have expressed disappointment.[17]

Impact on religion

COVID-19 has had a profound effect on the way religion is practised all over the world. There has been cancellations of worship services for various faiths, including the closure of temples, churches, mosques and synagogues, along with the cancellation of religious pilgrimages. The spread of coronavirus has been linked to congregation of churches in South Korea and Tablighi Jamaat in Southeast Asia and India. All places of worship, from China to Vatican City to Iran to India, have been closed as part of the efforts to prevent the virus from spreading further. According to *Time* magazine, Saudi officials urged Muslims to postpone the Hajj until the coronavirus pandemic has been brought under control. In February, the kingdom decided to close off the holy cities of Mecca and Medina to foreigners because of the virus, a step which wasn't taken during the devastating 1918 flu epidemic.[18]

Similarly, the Navaratri festival of the Hindus had quieter celebrations in India. The temples were empty, because people decided to celebrate the festival at home amid the ongoing lockdown. Ramnavami mela, usually attended by around 1.5 million devotees, was scrapped in Uttar Pradesh, a state in India. The majority of temples in India are under lockdown, and all kinds of religious gatherings have been banned.

Churches in many countries of the world are closed, and Masses have been postponed. The traditional Mass to celebrate Easter at St Peter's Basilica in Rome, followed by the pope's Urbi et Orbi blessing in St Peter's Square, is usually attended by tens of thousands of people from all over the world. The square was closed off by Italian police, and public Masses and funerals were cancelled. Pope Francis livestreamed Easter Sunday Mass from an empty St Peter's Basilica. He urged the European Union to show solidarity amid the coronavirus pandemic.[19] To limit the spread of COVID-19, bishops' conferences and dioceses around the world began suspending public Masses. The Diocese of London announced all Masses and services, including missions and devotions, would be suspended at least until April 30. All churches in the city were to be closed until then.[20] In addition, the diocese says funeral Masses and luncheons are not permitted. The Catholic Church in France suspended all public Sunday Masses, weddings and baptisms. Spain, the Netherlands, Belgium, Germany, Slovakia, Malta and Austria also suspended public Masses.[21]

In the US, from Seattle to Washington, DC., bishops are encouraging the faithful, especially those over 60 years old or afflicted with underlying health conditions, to stay home, fearing the spread of pandemic. The Archdioceses of Seattle, Little Rock, Salt Lake City and Santa Fe have decided to cancel public Masses. Major archdioceses, such as in Chicago and Newark, have undertaken similar precautious.[22]

Religious leaders have been repeatedly requesting worshippers to pray at home, to prevent contact and further spread of infection. Relief wings of almost all religious organizations are doing their best to fight against the pandemic by providing materials like masks, sanitizer and medical equipment, along with donating money and providing food to the poor and destitute.

Regardless of the outcome of this current pandemic, it is going to change the way religion is practised the world over.

Impact on common people

Besides these major impacts, people, especially the poor and migrants, have been badly affected by COVID-19. The major modes of transport like trains, buses and airports have been closed down. This has affected many poor and homeless all over the world. With cities under lockdown, their survival becomes much more difficult. The migrant workers in India have started leaving the cities, defying the orders of lockdown, and have started going back to their villages.[23] They have been forced to walk miles without knowing where their next meal will come from.

People have been stuck at home in the cities under lockdown. Social events like marriages have been postponed; several families have been separated, even stranded; and people are not able to look after their sick relatives.

There has been a radical shift in work practices, with many nonessential jobs witnessing a move to remote work overnight. Working from home is perhaps the new normal. Even students are attending their classes online. Technology has connected the global community like never before – the smartphone becoming the window to the world. Surely in future, the digital world will completely take over the analogue world, and the COVID-19 pandemic is just the beginning.

Famous personalities affected by COVID-19

From common people to famous personalities, coronavirus has spared no one. Worldwide, more than 2 million people have been affected by COVID-19, including leading politicians and personalities from the entertainment and sports worlds.

Leading politicians

- Prince Charles, the eldest son and heir to Queen Elizabeth II, tested positive for COVID-19 and went into self-isolation in Scotland.
- British Prime Minister Boris Johnson tested positive for COVID-19, and after staying in an ICU for three nights, he has been discharged.
- UK Health Minister Nadine Dorries was the first British politician to be diagnosed with COVID-19.
- Canadian Prime Minister Justin Trudeau's wife, Sophie Grégoire Trudeau, also tested positive for COVID-19.
- Australia's Home Affairs Minister Peter Dutton said in a statement that he was diagnosed with COVID-19.
- Iranian Deputy Health Minister Iraj Harirchi confirmed having tested positive for COVID-19.

Entertainment personalities

- Veteran actor Tom Hanks and his wife, Rita Wilson, were two of the first public figures to be diagnosed with COVID-19.
- *Game of Thrones* and *The Fast of the Furious* actor Kristofer Hivju, who played the role of Tormund in the popular HBO series, also tested positive for COVID-19.
- Indira Varma, another star from *Game of Thrones*, tested positive for COVID-19.
- Actor Idris Elba is among those who have tested positive for COVID-19.
- Spanish opera singer Placido Domingo also tested positive for COVID-19.
- Rachel Matthews, who leant her voice to Honeymaren in *Frozen II*, has tested positive for COVID-19.
- Kanika Kapoor, the Bollywood singer, was the first celebrity from India to have tested positive for COVID-19.

Sports personalities

- Arsenal manager Mikel Arteta tested positive for COVID-19, after which the Gunners' Premier League match at Brighton was postponed.
- Juventus and Italy defender Daniele Rugani was affected by COVID-19.
- Basketball player Kevin Durant was among the four basketball players from the Brooklyn Nets who tested positive for COVID-19.
- Utah Jazz centre Rudy Gobert and guard Donovan Mitchell had COVID-19 but have since been cleared.
- Chelsea footballer Callum Hudson-Odoi was the first Premier League player to be diagnosed with COVID-19.
- Other athletes affected by COVID-19 include Paulo Dybala, Ezequiel Garay, Albin Ekdal, Manolo Gabbiadini and Omar Colley.

Impact of COVID-19 on the global economy

The economic consequence of a pandemic like COVID-19 is quite significant. It has been noticed in the past that even when the health impact of an outbreak remains limited, its economic impact can be substantially high. For instance, a paper by IMF highlights that the GDP growth of Liberia witnessed a decline of eight percentage points from 2013 to 2014 during the Ebola outbreak in West Africa, while the country's overall death rate during the same period actually decreased.[24]

There are multiple ways that an economy bears the brunt of such a pandemic. First and foremost, such a health crisis necessitates substantial increase in public expenditure on ramping up the healthcare system that is needed to treat the infected and control the outbreak. For instance, special isolation wards are being created in public hospitals, and free treatment is offered to COVID-19-affected patients, with the cost being borne by governments.

In special cases, health insurance packages are offered to healthcare workers, as is the case in India, where the government has announced insurance coverage of 5 million rupees per person for three months to frontline healthcare workers – including sanitation staff, paramedics and nurses, and other special category workers and doctors – who are at the forefront to tackle COVID-19 and who face the highest risk of contracting the disease. Rising cases of coronavirus infection also increase the cost of health insurers. As per the estimates of the S&P Global, an international credit rating organization, a severe coronavirus pandemic in the US could cost the country's health insurers about US$90 billion in medical claims.[25]

Second, the precautionary measures taken across countries to contain the spread of a pandemic severely impacts economic activity, which comes to a grinding halt during the outbreak period. Social-distancing efforts lead to closure of schools, offices, commercial establishments, factories, transportation and public services. This disrupts economic and social activities.

In the case of COVID-19, which started in China, the immediate impact was felt in the global supply chain, which faced major disruption as many factories in the Hubei province – the epicentre of the coronavirus outbreak in the country – were shut down, resulting in decline in China's industrial production and a subsequent delay in the supply of goods from China to other parts of the world. Most factories in the affected areas remained closed for the better part of January and February. Even in March, the production units worked much below their capacity levels.

As a result, China's industrial output declined sharply by 13.5% in the months of January and February 2020 compared to the same period of the previous year. This had been the weakest growth recorded since January 1990.[26] The industrial production continued to decrease in March, with output registering a decline of 1.1%. Consequently, for the first quarter of 2020, the country's overall industrial production declined by 8.4%.[27] China's exports also contracted during the first three months of 2020, with overseas shipments falling 17.2% in January and February and by 6.6% in March compared to the same months a year earlier.[28]

China is a major global supplier of products and components to the world's leading manufacturing companies, with nearly 20% of global trade in the manufacturing of intermediate products originating in China, according to the UN's trade and development agency, the United Nations Conference for Trade and Development (UNCTAD).[29] Thus, a slowdown in manufacturing activity in China and a decline in exports directly affected industrial activity in other countries which source critical inputs or final products from China.

Sectors having global supply chain linkages such as automobiles, electronics and chemical products have faced imminent raw material and component shortages, which hampered their production schedules. Moreover, with the coronavirus gradually spreading to most other countries, including major European and Asian countries and the US, the manufacturing activity in these countries also started suffering on account of similar factory closures or poor attendance by factory workers, as happened in China. In fact, in many countries which observed complete lockdown, economic activity in such countries came to a near halt.

Third, the slowdown in manufacturing activity also affected the growth of international trade as countries faced difficulty in exporting or importing goods across borders. UNCTAD has estimated that a 2% reduction in exports from China could drag down exports across global value chains by US$50 billion during 2020.[30]

Fourth, under different scenarios of the spread of the pandemic, there is a possibility of a 5%–15% decline in global foreign direct investment (FDI) flows from the level predicted earlier. The countries that have been most severely affected by the pandemic will see a relatively more adverse impact on their FDI flows than will other countries, UNCTAD has cautioned. There has already been a drop in FDI flows into China, which decreased by 10.8% in January–March 2020 compared to the same period the previous year.[31]

Fifth, demand for many goods and services have declined sharply during the coronavirus outbreak. For instance, countries' closing international borders to contain the spread of the pandemic has directly affected demand in some sectors, like travel, tourism and hospitality, which in many countries have reached the brink of bankruptcy. According to the International Air Transport Association, airlines could lose, in passenger revenues, up to US$113 billion if the coronavirus continues to spread further. According to the Mobility Market Outlook on COVID-19, as compared to 2019, the global revenue for the travel and tourism industry is estimated to decrease by around 17% in 2020 to US$568.6 billion.[32] The revenue for the travel and tourism industry in Asia could be the most affected by the pandemic.

Similarly, the consumption of some essential and discretionary items has been hit badly due to closing of marketplaces, shopping complexes and movie halls, which either have been closed in some countries or have had reduced attendance. The retail sector has been badly hit as a result of this scenario and may lead to large-scale layoffs.

Entertainment and sports are the other two sectors which are experiencing a similar fallout. The outbreak of coronavirus has impacted the world of live events, with multiple sports and promotional events having been either cancelled or postponed. Event organizers are estimated to be incurring huge losses due to the cancellation of events.

Sixth, subdued investors' confidence levels due to greater uncertainty about the future course and repercussions of COVID-19 made the financial markets extremely volatile worldwide during the early months of 2020, leading to huge market crashes and wealth erosion, which in turn adversely impacted consumption levels. Stock markets all over the world reportedly lost about US$5 trillion in wealth during the first week of March alone, as measured by the MSCI all-country index.[33]

Assessing the economic impact on the global economy

Because of the presence of several interplaying factors, assessing the actual magnitude of economic losses due to the pandemic becomes extremely difficult. However, there is strong agreement among leading economists that the COVID-19 pandemic will have a severe adverse impact on global economic growth. Because China is the second largest economy in the world, contributing about 18%–19% to the world GDP[34] and as a major trade partner for many countries, the slowdown in China's economy is expected to have an adverse impact on the growth of the global economy.

Some of the leading global financial institutions and policy think tanks have attempted to estimate the likely economic impact of the novel coronavirus on global economic growth. Many of the analysts have attempted to draw analogies with similar health outbreaks from the past, like SARS in 2003, which resulted in a decline of 0.5% to 1% in China's growth that year and the global economy suffered a cost of about US$40 billion (or 0.1% of global GDP). However, most organizations, like the Organisation for Economic Co-operation and Development (OECD), are of the opinion that since the global economy today has become significantly more interconnected than it was in 2003 and since China has assumed a greater role in global output, trade, tourism and commodity markets, the economic impact of China's slowdown on other economies will be higher this time.[35] Besides the impact of slowdown of China's economy, growth of individual countries has also been severely impacted due to the slowdown observed in the economic activity of these countries.

The coronavirus could also make a deeper impact this time because the global economy was already going through a slow phase before the COVID-19 outbreak. For instance, China was growing at a rate of 6%, which is its lowest growth observed since 1990.[36] Therefore, a crisis of this magnitude can translate into a bigger slump for the world economy at this juncture.

UNCTAD had warned in its Trade and Development Report update released in March 2020 that many countries in the world would experience recession. The agency had anticipated a slowdown in the global economy to under 2% in 2020, which could lead to a global income decline of about US$1 trillion. This cost could even go up to US$2 trillion in the case that the world economy were to grow by a meagre 0.5%.[37] Similarly, the OECD, comprising 36 member countries, had projected that an intensive and long-lasting coronavirus outbreak which spreads widely throughout the Asia Pacific region, Europe and North America could bring the global growth down to 1.5% in 2020.[38] However, within a month's time, i.e. by April 2020, even these estimates have started looking optimistic given the ferocity of the COVID-19 pandemic, bringing more and more economies under its grip with each passing day. This has prompted all organizations to relook at their projections, and many have come out with a revised growth projection for the world economy.

The International Monetary Fund (IMF) in its world economic outlook released in April 2020 has in fact drawn a completely gloomy picture for the world economy for the current year. The IMF has projected that as a result of the pandemic, the global economy will actually contract sharply by 3% in 2020 (Figure 7.1). The present crisis has been determined to be much worse than the 2008–2009 financial crisis. The US economy is projected to shrink by 5.9%, and the euro area by 7.5%; meanwhile, India and China will grow at a much lower rate of 1.9% and 1.2% respectively.[39]

The latest OECD estimates suggest that unprecedented measures like the lockdown of cities are directly resulting in a sharp reduction in output, household spending, corporate investment and international trade. It is also estimated that for each month of strict lockdown or containment measures, there can be a loss of two percentage points in annual GDP growth, and in the case that the shutdown

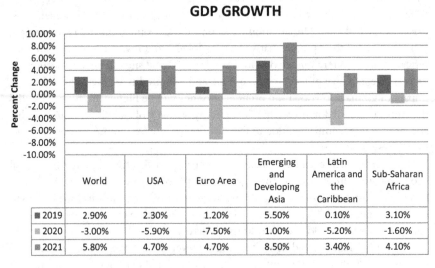

GDP GROWTH

	World	USA	Euro Area	Emerging and Developing Asia	Latin America and the Caribbean	Sub-Saharan Africa
2019	2.90%	2.30%	1.20%	5.50%	0.10%	3.10%
2020	-3.00%	-5.90%	-7.50%	1.00%	-5.20%	-1.60%
2021	5.80%	4.70%	4.70%	8.50%	3.40%	4.10%

FIGURE 7.1 Growth Projection by IMF

Source: IMF.

extends for three months, with no offsetting factors, the GDP growth could come down by four to six percentage points from the level that could have been attained under normal circumstances.

Impact on jobs

The partial or complete lockdown observed in various countries has led to the shutdown of factories and offices. This has caused severe economic disruption, with businesses across several sectors taking significant revenue losses. This has made the people working with these sectors vulnerable to income as well as job losses. According to International Labour Organisation (ILO), nearly 2.7 billion workers, representing about 81% of the world's total workforce, have been affected as a result of the economic fallout of the COVID-19 pandemic.[40]

The ILO Monitor[41] highlights that in many countries, large-scale employment contraction has already started, and the decline in employment will affect the number of jobs and the aggregate hours of work available. People working in the most-affected sectors, like retail trade, travel and tourism and manufacturing, are at a higher risk of job loss. According to ILO estimates, about 1.25 billion workers, representing almost 38% of the global workforce, are employed in these worst-hit sectors. The situation is also particularly bad for nearly 2 billion people who are working in the informal sector. The people working informally, mainly in developing countries, do not have access to the basic protections of formal jobs. This makes them most vulnerable to job cuts. The ILO also cautions that the ultimate increase in global unemployment in 2020 will depend largely on the pace at which the economies recover in the second half of the year and how effectively policy measures are deployed to boost labour demand.

Increase in poverty

According to a World Bank blog post,[42] the COVID-19 pandemic could push about 40–60 million people into extreme poverty. This estimate has been made based on the analysis of household survey data and growth projections of 166 countries. As per the projections of the IMF, emerging and developing economies will contract only by 1%, and advanced economies will contract by 6%. However, the research points out that since more people live close to the international poverty line in developing and emerging countries, they will experience the highest increase in extreme poverty. Nearly 23 million people will be pushed to poverty in sub-Saharan Africa and 16 million in South Asia.

Three countries in particular are estimated to see the highest change in the number of poor, including India (12 million), Nigeria (5 million) and the Democratic Republic of Congo (2 million). More than 1 million people in countries such as Indonesia, South Africa and China will also be pushed into extreme poverty as a consequence of COVID-19.

Impact on food availability

The transportation and processing of food and food products have been affected due to restrictions imposed on the movement of people and goods as a precautionary measure to contain the COVID-19 pandemic. This has made it difficult for people to get basic food items on time. The UN Food Relief Agency has issued a warning: as the world is fighting against the coronavirus pandemic, it is also 'on the brink of a hunger pandemic', which could lead to 'multiple famines of biblical proportions' within a few months if immediate action is not taken.[43] As per the analysis of the UN World Food Program, about 130 million people could face starvation-like situations by the end of 2020 as a result of COVID-19. This is in addition to nearly 821 million people who go to bed hungry every night all over the world, and another 135 million people are facing crisis levels of hunger due to other reasons.

Need for policy measures

The COVID-19 pandemic is clearly not just a major health crisis that the world is facing; it is affecting the global economy in a number of other ways. Therefore, while it is important to take all possible measures to strengthen healthcare facilities and contain the pandemic, it is equally important to emphasize minimizing the economic impact of the outbreak and put economies back on track.

Given the adverse consequences of the pandemic on the economic health of countries as well as the overall wellbeing of people, policymakers need to devise and implement appropriate measures to provide adequate support to the ailing businesses while stimulating demand and job creation in the economy. Specific targeted fiscal and monetary packages need to be devised for the worst-affected sectors and small businesses. This is essential to save economies from the worst effects of the COVID-19 pandemic.

The severe economic slowdown faced by COVID-19-affected countries have prompted many of them to come out with special stimulus packages to revive their economies. Most of the countries have attempted to extend support directly by increasing government expenditure by way of providing funds to small firms, enabling them to sustain their businesses, keeping people employed, giving specific fiscal packages to the worst-hit sectors (such as aviation, tourism etc.), increasing fund allocation for infrastructure development and more. Some governments have announced the cancellation of certain taxes, thereby taking a direct hit on revenue collection for the year.

In countries like India, money has also been directly transferred in the hands of the most vulnerable sections of the economy. Under a 1.7 trillion rupee (US$22.6 billion) relief package announced in March 2020, the Indian government promised to provide gas cylinders to 80 million poor families free of cost, pay 1000 rupees to 30 million senior citizens and 500 rupees to over 200 million women per month for a period of three months.

Many governments have also decided to defer some payments, including term loans, interests, taxes and utility bills. This has helped lower the immediate financial burden of borrowers and taxpayers. Direct liquidity has also been injected into the system through specific credit lines to small businesses and financial institutions. Such measures help by improving the cash flow or the liquidity position of firms.

Among all the countries that have announced fiscal packages so far, the US has announced the largest package in financial terms, worth US$2.7 trillion, amounting to nearly 13% of its GDP. However, this is not the highest as a proportion to GDP. Japan has in fact decided to spend about 21% of its GDP, worth US$1.1 trillion, to help people and businesses come out of the COVID-19 crisis.[44] Governments across the European Union have also started announcing and implementing large-scale fiscal measures to control the economic consequences.[45] Spain and Italy, the two most-affected countries in Europe, have also announced stimulus packages estimated to be around 7.3% and 5.7% of GDP respectively, until May 10, 2020.[46] The stimulus packages announced by the top ten countries, as a proportion to their GDP, are presented in Figure 7.2.[47]

Recognizing the COVID-19 pandemic as an economic emergency, the UK government announced a fresh £330 billion (US$398 billion) of government-backed loans and guaranteed on May 17, 2020, to help businesses impacted by the coronavirus outbreak. As part of the package, smaller businesses will be able to access cash grants of up to £25,000 to minimize the economic impact of COVID-19, and mortgage providers will be able to provide a three-month mortgage holiday to those who are in need.[48]

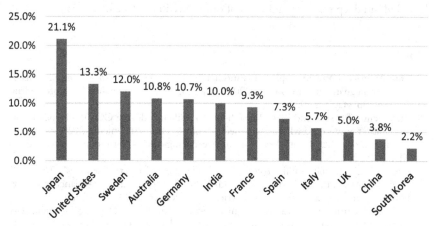

Country-Specific Stimulus Package Comparison (as Share of GDP)

FIGURE 7.2   Country-Specific Stimulus Packages

Source: Ceyhun Elgin.

Note: As of May 10, 2020; India announced an economic stimulus package on May 12, 2020.

India too has announced a massive 20.97 trillion rupee (about US$265 billion) stimulus package for its economy, comprising nearly 10% of the country's GDP. The package includes a programme funded by 2.3 trillion rupees for cheap loans for farmers to revive the rural economy and several steps for easing the hardship of the urban poor who have been hit by the lockdown;[49]8.01 trillion rupees of liquidity measures announced by the Central Bank; a 5.94 trillion rupee package for extending credit lines to small businesses and support to shadow banks and electricity distribution companies; 3.10 trillion rupees for free food grain to stranded migrant workers and credit to farmers; 1.5 trillion rupees for agriculture infrastructure and other measures for agriculture and allied sectors; a 1.92 trillion rupee package of free food grain and cooking gas to the poor and cash to some sections; and around 481 billion rupees for structural reforms.[50]

Although countries are taking measures at individual levels to tackle the situation on their respective turfs, this battle should also be fought by all economies together with mutual cooperation, helping each other mitigate the negative impact of the pandemic. This is critical for an early revival of the global economy.

The enormous impact of the COVID-19 pandemic on society, religion, economies and mental health will take time to recover from. This 21st-century pandemic will certainly give us an opportunity to think about what went wrong with our lives and our lifestyles and wonder whether it could have been prevented or, if not, better managed.

Preventing infection plays a great role in controlling the COVID-19 pandemic. Every effort should be made to protect oneself from getting infected. It also is our duty to protect others from getting infected. Self-isolation, quarantine and social distancing are important methods of containing the COVID-19 pandemic. These non-therapeutic measures are the only means of control in the absence of a vaccine and effective treatments. The pandemic will decline in due course of time, but will leave behind deep scars to remind us of our frailty in the face of adversity.

Notes

1 36m Americans now unemployed as another 3m file for benefits, *The Guardian*, available at www.theguardian.com/business/2020/may/14/unemployment-us-data-coronavirus, accessed on May 16, 2020

2 Statement – Physical and mental health key to resilience during COVID-19 pandemic, available at www.euro.who.int/en/media-centre/sections/statements/2020/statement-physical-and-mental-health-key-to-resilience-during-covid-19-pandemic, accessed on March 28, 2020

3 German state finance minister commits suicide after 'virus crisis worries', *The Week*, available at www.theweek.in/news/world/2020/03/29/german-state-finance-minister-commits-suicide-after-virus-crisis-worries.html, accessed on May 16, 2020

4 Suspected coronavirus patient commits suicide at Delhi's Safdarjung Hospital, *Live Mint*, available at www.livemint.com/news/india/coronavirus-suspected-patient-commits-suicide-at-delhi-s-safdarjung-hospital-11584561374632.html, accessed on May 16, 2020

5 Wu P, Fang Y, Guan Z, et al., The psychological impact of the SARS epidemic on hospital employees in China: Exposure, risk perception, and altruistic acceptance of risk, *Can J Psychiatry*, May 2009; 54(5): 302–311

6 Wu KK, Chan SK, Ma TM, Posttraumatic stress after SARS, *Emerg Infect Dis*, August 2005; 11(8): 1297–1300

7 Centers for Disease Control and Prevention. CDC twenty four seven: Saving lives, protecting people, available at www.cdc.gov/coronavirus/2019-ncov/prepare/managing-stress-anxiety.html, accessed on March 28, 2020

8 Chen S, Yang J, Yang W, et al., COVID-19 control in China during mass population movements at New Year, *Lancet*, March 7, 2020; 395, Published Online February 20, 2020, doi:10.1016/S0140-6736(20)30421-9

9 Coronavirus: Hong Kong Disneyland to be closed to help prevent spread of virus, *The Hindu*, available at www.thehindu.com/news/international/coronavirus-hong-kong-disneyland-to-be-closed-to-help-prevent-spread-of-virus/article30656912.ece, accessed on May 16, 2020

10 London Book Fair cancelled, available at www.londonbookfair.co.uk/, accessed on March 30, 2020

11 Geneva Motor Show cancelled after coronavirus causes government to ban large events, available at www.theverge.com/2020/2/28/21156368/geneva-motor-show-2020-canceled-coronavirus-concerns, accessed on March 30, 2020

12 Joint Statement from the International Olympic Committee and the Tokyo 2020 Organising Committee, available at www.olympic.org/news/joint-statement-from-the-international-olympic-committee-and-the-tokyo-2020-organising-committee, accessed on March 30, 2020

13 Euro 2020 postponed until 2021 due to coronavirus, available at www.espn.in/football/uefa-european-championship/story/4074294/euro-2020-postponed-until-2021-due-to-coronavirus-sources, accessed on March 30, 2020

14 Premier League 2020–21 season won't start until this one ends – sources, available at www.espn.in/football/english-premier-league/story/4076278/coronavirus-premier-league-extends-suspension-until-april-30, accessed on March 30, 2020

15 Coronavirus: Bundesliga and Ligue 1 suspend fixtures, follow Premier League, La Liga and Serie A's suit, available at www.cbssports.com/soccer/news/coronavirus-bundes liga-and-ligue-1-suspend-fixtures-follow-premier-league-la-liga-and-serie-as-suit/, accessed on March 30, 2020

16 List of all the cricket series affected by coronavirus: Full coverage, available at www.espncricinfo.com/story/_/id/28898417/full-coverage, accessed on March 30, 2020

17 Wimbledon cancelled for first time since WWII over coronavirus, available at www.france24.com/en/20200402-wimbledon-cancelled-for-first-time-since-wwii-over-coronavirus, accessed on April 4, 2020

18 https://time.com/5813644/saudi-arabia-hajj-postpone-coronavirus/, accessed on April 6, 2020

19 Pope urges EU to show solidarity amid coronavirus crisis in Easter message, *The Guardian*, available at www.theguardian.com/world/2020/apr/12/pope-francis-urges-eu-to-show-solidarity-amid-coronavirus-crisis-in-easter-message, accessed on May 16, 2020

20 Coronavirus: Churches closed, masses suspended until May in London, Ont., available at https://boom1019.com/news/6696610/church-mass-london-coronavirus/, accessed on May 16, 2020

21 Update: Churches begin canceling masses in effort to stem COVID-19 pandemic, available at www.catholicnews.com/services/englishnews/2020/churches-begin-canceling-masses-in-effort-to-stem-covid-19-pandemic.cfm, accessed on May 16, 2020

22 www.americamagazine.org/faith/2020/03/13/us-dioceses-suspend-masses-and-close-churches-coronavirus-pandemic-escalates, accessed on April 6, 2020

23 Explainer: Migrant crisis amid the coronavirus pandemic, *The Week*, available at www.theweek.in/news/india/2020/03/29/explainer-the-migrant-crisis-amid-the-coronavi rus-pandemic.html, accessed on March 30, 2020

24 Bloom DE, Cadarette D, Sevilla JP, Epidemics and economics, available at www.imf.org/external/pubs/ft/fandd/2018/06/economic-risks-and-impacts-of-epidemics/bloom.htm, accessed on March 30, 2020

25 Coombs B, S&P estimates a severe coronavirus outbreak in US could cost $90 billion in insured medical expenses, *CNBC*, available at www.cnbc.com/2020/03/13/sp-estimates-the-new-coronavirus-could-cost-us-insurers-90-billion-in-medical-expenses.html, accessed on March 30, 2020

26 China January–February industrial output shrinks 13.5%: Investment plunges 24.5%, available at www.cnbc.com/2020/03/16/china-january-february-industrial-output-drops-investment-plunges.html, accessed on March 17, 2020

27 China says its economy shrank by 6.8% in the first quarter as the country battled coronavirus, available at www.cnbc.com/2020/04/17/china-economy-beijing-contracted-in-q1-2020-gdp-amid-coronavirus.html, accessed on May 17, 2020

28 China January–February exports tumble, imports slow as coronavirus batters trade and business, available at www.cnbc.com/2020/03/07/coronavirus-china-jan-feb-2020-trade-data-reflects-outbreaks-impact.html and www.cnbc.com/2020/04/14/china-reports-march-2020-trade-exports-imports-data-amid-coronavirus.html, accessed on May 17, 2020

29 Global trade impact of the coronavirus (COVID-19) epidemic, UNCTAD, available at https://unctad.org/en/PublicationsLibrary/ditcinf2020d1.pdf, accessed on March 30, 2020

30 Coronavirus outbreak has cost global value chains $50 billion in exports, UNTCAD, available at https://unctad.org/en/pages/newsdetails.aspx?OriginalVersionID=2297, accessed on May 16, 2020

31 China says has not seen large-scale exodus of foreign capital amid coronavirus, available at https://in.reuters.com/article/china-economy-foreign-investment/china-says-has-not-seen-large-scale-exodus-of-foreign-capital-amid-coronavirus-idINKCN21Y11U, accessed on May 17, 2020

32 Forecasted change in revenue from the travel and tourism industry due to the coronavirus (COVID-19) pandemic worldwide from 2019 to 2020, available at www.statista.com/forecasts/1103426/covid-19-revenue-travel-tourism-industry-forecast, accessed on March 30, 2020

33 Coronavirus may knock 1.3 per cent off global GDP growth, *National Herald*, available at www.nationalheraldindia.com/opinion/coronavirus-may-knock-13-per-cent-off-global-gdp-growth, accessed on March 30, 2020

34 China: Share of global gross domestic product (GDP) adjusted for purchasing-power-parity (PPP) from 2012 to 2024, available at www.statista.com/statistics/270439/chinas-share-of-global-gross-domestic-product-gdp/, accessed on May 17, 2020

35 OECD Interim Economic Assessment – Coronavirus: The world economy at risk, available at www.oecd.org/berlin/publikationen/Interim-Economic-Assessment-2-March-2020.pdf, accessed on May 17, 2020

36 Vinelli A, Weller CE, Vijay D, The economic impact of coronavirus in the U.S. and possible economic policy responses, available at www.americanprogress.org/issues/economy/news/2020/03/06/481394/economic-impact-coronavirus-united-states-possible-economic-policy-responses/, accessed on March 30, 2020

37 This is how much the coronavirus will cost the world's economy, according to the UN, World Economic Forum, available at www.weforum.org/agenda/2020/03/coronavirus-covid-19-cost-economy-2020-un-trade-economics-pandemic/, accessed on March 30, 2020

38 OECD Interim Economic Assessment – Coronavirus: The world economy at risk

39 World Economic Outlook, April 2020: The great lockdown, available at www.imf.org/en/Publications/WEO/Issues/2020/04/14/weo-april-2020, accessed on May 17, 2020

40 ILO monitor: COVID-19 and the world of work. Second edition, Updated estimates and analysis, ILO, available at www.ilo.org/wcmsp5/groups/public/–dgreports/–dcomm/documents/briefingnote/wcms_740877.pdf, accessed on April 22, 2020

41 Ibid.

42 Mahler DG, Lakner C, Castaneda Aguilar RA, et al., The impact of COVID-19 (Coronavirus) on global poverty: Why Sub-Saharan Africa might be the region hardest hit,

World Bank, available at https://blogs.worldbank.org/opendata/impact-covid-19-coronavirus-global-poverty-why-sub-saharan-africa-might-be-region-hardest, accessed on April 22, 2020

43 UN warns of 'biblical' famine due to COVID-19 pandemic, *France 24*, available at www.france24.com/en/20200422-un-says-food-shortages-due-to-covid-19-pandemic-could-lead-to-humanitarian-catastrophe, accessed on April 22, 2020

44 How global coronavirus stimulus packages compare, available at www.forbes.com/sites/niallmccarthy/2020/05/11/how-global-coronavirus-stimulus-packages-compare-infographic/#9e94b60ca52c, accessed on May 18, 2020; How does Modi's COVID-19 package compare to spending by other countries? Available at www.theweek.in/news/biz-tech/2020/05/12/how-does-modis-economic-package-compare-to-spending-by-other-countries.html, accessed on May 18, 2020

45 The fiscal response to the economic fallout from the coronavirus, available at www.bruegel.org/publications/datasets/covid-national-dataset/, accessed on May 18, 2020

46 How global coronavirus stimulus packages compare

47 Ibid.; How does Modi's COVID-19 package compare to spending by other countries?

48 'Whatever it takes': UK pledges almost $400 billion to help businesses through coronavirus, available at www.cnbc.com/2020/03/17/uk-announces-massive-aid-package-for-coronavirus-hit-industries.html, accessed on May 18, 2020

49 Govt unveils ₹2.46 tn cheap loans plan to revive farm economy, available at www.livemint.com/news/india/govt-unveils-2-46-tn-cheap-loans-plan-to-revive-farm-economy-11589482005015.html, accessed on May 18, 2020

50 Economic stimulus includes Rs 8 trillion liquidity measures by RBI: FM, available at www.business-standard.com/article/economy-policy/economic-stimulus-includes-rs-8-trillion-liquidity-measures-by-rbi-fm-120051700379_1.html, accessed on May 18, 2020

8

CONTAINING THE CONTAGION

One of the most important aspects of mitigating and containing a pandemic is situational awareness, which itself means having accurate knowledge of the threat of the ongoing coronavirus pandemic and the available resources to fight it. It requires complete coordination between healthcare workers, public health authorities, diagnostic infrastructure, communication systems, policymakers, the judiciary, law enforcement agencies and other government agencies at various levels of governance. Policy decisions are required to detect and diagnose the disease by appropriate testing, often emphasized by the WHO, find the areas where disease transmission is going on and decide the proper allocation of resources. Case detection, contact tracing and surveillance are the most essential initial steps in containing the COVID-19 pandemic. Disease surveillance, the backbone of public health, provides data needed to understand the pandemic threat and give early warnings and help in formulating a targeted response. Countries with more-robust public health systems could manage and control the spread and damage caused by the COVID-19 pandemic. However, there is no guarantee that it will be sufficient to stop the ongoing spread of COVID-19.

A lack of appropriate diagnostic kits, in terms of both quality and quantity, has proved to be a great hindrance in many countries. Early detection followed by isolation of affected patients plays the most crucial role in further containing the spread. The lack of antiviral medicine against SARS-CoV-2, the lack of vaccine and overwhelmed healthcare infrastructure mean nonpharmaceutical measures are the main modalities of pandemic containment. In the absence of therapeutic countermeasures, public health interventions play a greater role in disease mitigation and containment. The public health measures that has been employed to tackle the COVID-19 pandemic range from relatively simple techniques like disease surveillance and hygiene measures to more-restrictive interventions like social distancing, self-isolation travel restrictions, quarantine and lockdowns of cities or entire countries.

DOI: 10.4324/9781003345091-8

The pandemic is still spreading at an alarming rate. It is now the responsibility of everyone in the world to take appropriate steps to stop its spread and contain the pandemic. One must not leave everything to governments and healthcare workers. It is a war that needs to be fought together by everybody at every level.

Prevention of infection

Know your enemy well

Before one can fight this invisible enemy, one has to know what to fight and how to fight. So the first step is to learn about COVID-19, and the second is to learn how to prevent infection. There is an avalanche of information about coronavirus in the media. But it is important to differentiate between genuine information and false information. Social media is doing more harm than good by acting as a medium for spreading misinformation and fake news. Few reliable books are available. Literature published in reputed medical journals is beyond the comprehension of common people. So the most reliable sources of information are those published by the WHO[1] and the CDC,[2] which are readily available on their respective websites. They are written in plain language so that everybody can understand them.

After gaining knowledge, the next step is to understand how the infection can be prevented from spreading from person to person. The new coronavirus spreads by droplet spread and direct contact with a COVID-19 patient. Hygienic measures have been used to prevent respiratory infections, and they have been put into use both during the Spanish flu and SARS. These include the use of face masks, gloves, hand washing and proper etiquette for coughing, sneezing and spitting. However, at the same time, it is important to inform the general public about how to use these measures, because misinformation through various means and especially social media results in people purchasing ineffective and expensive products.

Personal protection measures

How can an individual protect themselves and their family members from getting the new coronavirus? Engage in the following measures for practising good hygiene.

Hand washing

Wash your hands frequently, using soap and running water for 20 seconds. If your hands are not visibly dirty, then frequently clean them using alcohol-based hand rub. Do not touch your eyes, nose and mouth with unclean hands.

Clean your hands properly, especially in the following situations:

- After sneezing or coughing
- After coming home from outside or meeting other people, especially if they are ill

- After caring for the sick
- Before preparing food, eating or feeding children
- Before and after using the toilet, cleaning and so on
- When hands are visibly dirty
- Before and after changing diapers
- After handling animals or animal waste (pets also develop coronavirus infections)

Steps of hand washing

COVID-19 can survive for even two to three days on plastic and steel surfaces.[3] So there is a high possibility of touching contaminated surfaces and thereby getting infected promptly. To eliminate all traces of the virus on your hands, a quick scrub and a rinse won't cut it. What follows is a step-by-step process for effective hand washing.[4]

- Step 1: Wet hands with running water.
- Step 2: Apply enough soap to cover wet hands.
- Step 3: Scrub all surfaces of the hands – including back of hands, between fingers and under nails – for at least 20 seconds.
- Step 4: Rinse thoroughly with running water.
- Step 5: Dry hands with a clean cloth or single-use towel.

Hot water is not required for washing hands. Wash for at least 20 to 30 seconds. For hand sanitizer, use a sanitizer that contains at least 60% alcohol and rub it into your hands for at least 20 seconds. CDC recommends the use of alcohol based hand rub with more than 60% ethanol or 70% isoprpanol in healthcare settings.[5] Use paper towels or clean clothes to completely dry your hands.

Face masks

Face masks are effective at preventing coronavirus infection. Since the virus spreads by droplets from the mouth and nose, masks, which cover these areas, are extremely beneficial when you are near a person who has a cough and fever. However, face masks are in short supply, and local authorities are sharing guidelines for their use. So one must follow the guidelines given by health authorities of their respective countries. However, the WHO has given certain guidelines, which are equally important.[6] One should understand that one is not fully protected by using masks alone. One has to follow other precautions as well. Masks may induce a sense of false security. Similarly, using masks incorrectly will neither provide adequate protection nor prevent spreading the virus to others. A medical mask is not required for people who are not sick. If masks are used, best practices should be followed about how to wear, remove and dispose of them and for hand hygiene after removal. Medical masks are mandatory for those who have cough, fever and breathlessness. It must be worn by healthy people who are looking after a person suspected to have contracted SARS-CoV-2. It is also mandatory for healthcare workers who come into contact with patients outdoors, indoors and in ICUs.

Mask management

Mask management is extremely important. They are helpful if used properly; otherwise, the mask itself may turn out to be a source of infection. Use the following information to correctly wear a face mask:

- Before wearing a mask, wash your hands properly.
- Place the mask carefully, covering your nose and mouth without leaving any gap between the mask and face. Tie it tightly.
- Do not touch the mask with your hands.
- While removing it, do not hold the front part of the mask. Untie it from behind.
- After removing the mask, dispose of it immediately in a closed bin.
- Wash your hands with soap and water after you remove the mask.
- Do not reuse a single-use mask.
- Replace the mask if it becomes damp.

Whatever the guidelines, if there is no shortage of face masks, they should be used by everybody who is going out of their house, especially in a crowded place. Although airborne transmission has not yet been proven, further research may contradict the present finding. Besides, COVID-19 also coincides with flu season in most countries. Study the accompanying flowchart (Figure 8.1) regarding the health tips for COVID-19 prevention.

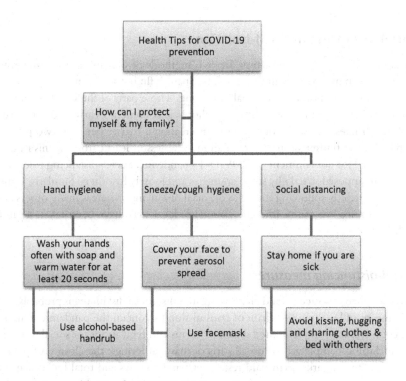

FIGURE 8.1   Health Tips for COVID-19 Prevention

Household disinfectant

It is crucial to frequently disinfect household objects during a pandemic. Repeatedly touched objects like door handles, knobs, electric switches, car doors and handles, tabletops and so on must be cleaned with disinfectant effective against the novel coronavirus. One need not buy costly disinfectant. It can be easily prepared at home.

Diluting your household bleach[7]

To make a bleach solution, mix the following:

• 5 tablespoons (1/3 cup) bleach per gallon of water

   OR

• 4 teaspoons bleach per quart of water

The bleach effective against coronavirus must contain 0.1% sodium hypochlorite. One must follow manufacturer's instructions for application and proper ventilation. Before preparing the solution, check the expiration date. Never mix household bleach with ammonia or any other cleanser. Alternatively, solution containing at least 60% alcohol may be used.

Social distancing and self-isolation

Social distancing and self-isolation are two methods of containing the pandemic that have been used widely in the past. The Spanish flu in the US in 1918 provided evidence of the importance of social distancing. Those parts of the US that banned public gatherings and closed theatres, schools and churches early had far lower peak death rates.[8] So social distance means creating a barrier between two people to prevent the transmission of a contagion. The lesson learnt from Spanish flu is now utilized for containment of COVID-19 pandemic. Social-distancing rules and norms are imposed strictly by some countries and loosely by others. Similarly, some people are strictly following the rules of social distancing, whereas many others are openly flouting the rules. In some countries, legal steps have been taken against such violators.

Social-distancing measures

In the absence of a vaccine and effective treatments, social distancing is probably the only means of breaking the chain of transmission to contain the pandemic. Different countries have implemented different norms and regulations for social distancing, including complete bans on all sorts of mass gatherings; the closure of public places like malls, parks, gyms and restaurants; and curfews and total lockdowns of cities and entire countries. People are forced to stay indoors. For a highly popu-lous country like India with a not-so-well-developed health infrastructure, entirely locking down India will definitely go a long way towards limiting the number of cases and casualties. Failure to ensure complete lockdown, which is not always easy, has most likely resulted in a huge number of cases and mortalities in countries like Italy and the US. However, there are many other causes for the same. As death tolls mount and fear rises, people will automatically shun public gatherings and volun-tarily isolate themselves.

Self-isolation norms

Self-isolation is a type of social distancing, but with a difference. It is a way of creating a barrier between an infected person and a healthy one. By isolating the infected person, the spread of the virus is limited. COVID-19 is extremely contagious, and one infected person is capable of infecting two people or prob-ably many more and the maximum incubation period of COVID-19 is 14 days. Hence, a minimum isolation period of 14 days is needed to prevent the spread of infection. If an infected person is not isolated, he will go on spreading to two or more others, and they in turn will further spread it to two more each. The infection will rise in an exponential manner and will soon get out of control. A person suspected to be infected should self-isolate at home in a separate, well-ventilated room. They should not go out at all and must depend entirely on their family members.

Flattening the curve

Another main aim of social distancing is flattening the curve of the epidemic. This means the virus will spread slowly in the community due to social distancing. This will reduce the peak number of cases during the epidemic. A flat curve means a smaller number of people are affected at any given time (Figure 8.2). If this curve isn't flattened, too many people will be affected at the same time, completely over-whelming the healthcare infrastructure, as has happened in parts of the US, Italy and China. If the curve is not flattened in a highly populous country like India, which has the potential to turn into the next epicentre of the pandemic, the conse-quences could be disastrous. Only time will tell whether the strong steps taken by Indian government will bear fruit or not. It is surprising to see that many countries are still dilly-dallying with the idea of complete lockdown for their country, even after observing the harsh realities of the pandemic in the US, Spain, Italy, China, France and Iran.

Once the pandemic has declined, it does not mean that it will completely disap-pear. During lockdown, the number of cases may decrease due to social distancing; however, the cases may start to rise after lifting lockdown measures. So govern-ments have to think carefully before lifting lockdown and need to prepare to reim-pose lockdown as the need arises.

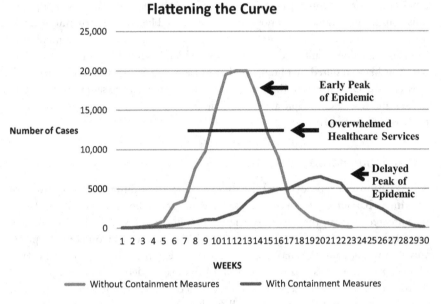

FIGURE 8.2 Flattening the Curve

Source: Author.

The rules of social distancing during the lockdown vary from country to country. However, in general, the following rules are in place in almost every country. Every individual must stay at home and can leave their respective homes only for the following reasons:

- For medical needs
- During emergency situations – medical or otherwise
- To buy essential food items and medicine
- To care for sick and vulnerable people

However, people involved in the fight against COVID-19, like doctors and other healthcare workers, sanitation workers, law enforcing personnel, and people involved in essential activities like water supply and sanitation, are allowed to leave home. When they leave, they must keep a minimum distance of 2 metres from others and maintain social distancing.

Shielding

In every society, there are vulnerable groups who must not go out during a pandemic. As much as possible, no one should visit them either. They are most prone to catching the virus. The NHS in the UK has urged people with serious health issues to not go out at all for at least 12 weeks.[9] This is known as shielding. These vulnerable groups are as follows:

- People who have cancer
- People taking steroids, which reduces immunity
- Organ transplant recipients
- People with certain genetic diseases
- People who have serious respiratory illness
- Pregnant women with heart disease

Protecting others from COVID-19

One must take care of oneself and take all possible precautions to prevent the spread of the novel coronavirus. At the same time, it is the prime responsibility of every person to protect others. One must try to comply with the following measures to protect others:

- While coughing or sneezing, cover your nose and mouth with a flexed elbow or a tissue rather than your palms.
- Throw the tissue into a closed bin immediately after use.
- Avoid close contact with others if you have a cough or fever.
- Do not spit in public.
- Seek early medical care if you are sick.
- Do not shake people's hands.

However, it is not easy for anyone to practise social distancing and self-isolation for long. Loneliness, emotional detachment, social and economic destruction, and infringements on civil rights and liberties are important consequences which cannot be ignored. The price one has to pay may be enormous. Similarly, the price of these measures may overwhelm the government. Developed countries might be able to manage the cost, but the developing world may find this difficult.

Quarantine

Quarantine is one of the oldest methods of controlling epidemics and pandemics. The practice of quarantine began in 14th-century Europe in response to the Black Death. During the epidemic of the Black Death, ships arriving in the ports were quarantined to prevent the spread of disease in the cities and towns. Quarantine centres were often created on nearby islands.

Quarantine means restricting the movement and activities of asymptomatic people who are exposed to a confirmed case of an infectious disease to prevent further spread. Establishing cordon sanitaire[10] is an essential element in the fight against COVID-19. Thousands of people have faced quarantine, and many more are going to face it during the COVID-19 pandemic. Some people are quarantined in their own home, whereas others are quarantined in a designated facility. Wherever they are kept, they are strictly isolated from a confirmed case of COVID-19. There are many who are not taking quarantine seriously and are trying their best to avoid it. Many have violated quarantine, and efforts are being made to trace them.

Quarantine is one of the most complex and controversial public health measures. At one end, it involves significant curtailments of individual civil liberties; at the other end, it is in the best interests of society. This throws up tensions between the rights of an individual and the enormous power of the government to curtail them. It can only be justified if the benefit to society outweighs the burden and harm to individuals.[11] A fine balance between the two is essential. Quarantine should not spark a tension between the citizen and the state.

Quarantine should be used only when careful consideration by public health specialists recommend its use. Experience is one of the best guides for this. Governments have the responsibility to make the decision to implement quarantine in a transparent manner. All these steps should be taken in advance before the virus reaches its shore. The transparency shown by the government will go a long way towards ensuring the public's acceptance of quarantine. The best quarantine is voluntary quarantine.

During quarantine, it becomes the responsibility of the government to provide adequate medical facilities and psychological supports, not just items for daily needs. Quarantine not only disrupts people's lives and liberties but involves a huge expenditure as well. The government should be in a position to bear the economic cost of quarantine and disruptions to economic activity. It should not pass on the cost to citizens.

Border control and international travel bans

An epidemic gets converted into a pandemic due to movement of people, animals and goods across international borders. Air travel is one of the most efficient means of spreading infection from one country to another, as has been observed in the COVID-19 pandemic. Governments have tried their best to close borders and impose travel bans to contain epidemics. However, such a massive disruption is not always possible. Public health law, especially the WHO's International Health Regulations,[12] provides a legal framework for governments across the world regarding travel bans and border closures.

The COVID-19 pandemic has also triggered travel bans and border closures in several countries to contain the pandemic. Some did it early, while some did it late in the course of the pandemic. Some countries, like India, shut down their borders completely and put a complete ban on airlines and foreign travel with few exceptions. Such disruptive measures may halt the march of the virus or may prove ineffective. Such transnational measures include screening at entry and exit, the collection and dissemination of passenger information, medical checkups at airports and quarantining and isolating sick passengers. As a result, passengers have faced significant harassment. The pandemic has disrupted travel plans, many have been stranded in foreign countries and airlines and tourism industries are going bankrupt. But it remains the responsibility of the national governments to implement transnational measures to halt the march of the pandemic within the limits of the rules set by International Health Regulations by the WHO.

The efficacy of preventive and containment measures depends on how well they are followed. Numerous issues come up, including the shortage of masks, gloves and sanitizer; social and psychological disruptions; economic crises; infringements on civil liberties; protecting the poor and vulnerable; and so on. Until an effective vaccine and medicines are discovered, these containment measures are our only hopes to prevent infection the spread of the COVID-19 pandemic.

One of the best methods of controlling a pandemic is a vaccine. However, a vaccine is not yet available. Many countries are trying their best to develop one. The WHO is coordinating with various stakeholders in their effort to develop an effective vaccine. COVID-19 has posed this global challenge which needs to be tackled urgently to contain this contagion. Besides this, COVID-19 has posed a few other challenges, like pandemic preparedness and developing good public health infrastructures all over the world. It will be prudent to understand these challenges and figure out a solution as soon as possible.

Notes

1 Coronavirus disease (COVID-19) advice for the public, WHO, available at www. who.int/emergencies/diseases/novel-coronavirus-2019/advice-for-public, accessed on March 30, 2020
2 How to protect yourself, CDC, available at www.cdc.gov/coronavirus/2019-ncov/ prevent-getting-sick/prevention.html?CDC_AA_refVal=https%3A%2F%2Fwww. cdc.gov%2Fcoronavirus%2F2019-ncov%2Fprepare%2Fprevention.html, accessed on March 30, 2020
3 How long can the coronavirus survive on surfaces or in the air? *Economic Times*, available at https://economictimes.indiatimes.com/magazines/panache/how-long-can-corona virus-live-on-surfaces-or-in-the-air/articleshow/74690737.cms?from=mdr, accessed on May 16, 2020
4 Everything you need to know about washing your hands to protect against coronavirus (COVID-19), UNICEF, available at www.unicef.org/coronavirus/everything-you-need-know-about-washing-your-hands-protect-against-coronavirus-covid-19, accessed on March 30, 2020
5 https://www.cdc.gov/coronavirus/2019-ncov/hcp/hand-hygiene.html
6 Advice on the use of masks in the community, during home care and in healthcare settings in the context of the novel coronavirus (COVID-19) outbreak, WHO, available at www.who.int/publications-detail/advice-on-the-use-of-masks-in-the-community-during-home-care-and-in-healthcare-settings-in-the-context-of-the-novel-coronavi rus-(2019-ncov)-outbreak, accessed on March 30, 2020
7 How to protect yourself, CDC

8   Beall A, Why social distancing might last for some time, *BBC*, available at www.bbc.com/future/article/20200324-covid-19-how-social-distancing-can-beat-coronavirus, accessed on March 30, 2020

9   Coronavirus: What are social distancing and self-isolation? *BBC*, available at www.bbc.com/news/uk-51506729, accessed on March 30, 2020

10   Salas-Vives P, Pujadas-Mora J-M, Cordons sanitaires and the rationalisation process in Southern Europe (nineteenth-century Majorca), *Med Hist*, July 2018; 62(3): 314–332

11   Cetron M, Landwirth J, Public health and ethical considerations in planning for quarantine, *Yale Journal of Biology and Medicine*, 2005; 78(5): 329–334; Ethical and legal considerations in mitigating pandemic disease: Workshop summary, available at www.ncbi.nlm.nih.gov/books/NBK54163/, accessed on March 30, 2020

12   International health regulations and epidemic control, WHO, available at www.who.int/trade/distance_learning/gpgh/gpgh8/en/index7.html, accessed on April 2, 2020

9

COVID-19

Global challenges

The COVID-19 pandemic has brought up crucial global challenges which need to be sorted out as soon as possible. In spite of several warnings, the world was not prepared to face a devastating pandemic. The public health infrastructure was not in place to deal with the health crisis triggered by the COVID-19 pandemic. COVID-19 has created critical challenges which if not taken care of early will result in further death and destruction not only in the current pandemic but also in a pandemic that the world may face in the future.

A COVID-19 vaccine

The first and most crucial challenge is developing a vaccine against COVID-19. Vaccines play the most important role in containing pandemics. They are the best therapeutic approach in preventing infection. Unfortunately, a vaccine for COVID-19 is not yet available. However, vaccine development programmes are taking place at an unprecedented speed and scale in various countries. More than 100 vaccine development programmes are going on at present, and fortunately, a few of them have reached an advanced stage of development within a short span of time. Genetic recombinant technology[1] has reduced the time for vaccine development by several years. Previously, on average, it used to take at least a decade to develop a safe and effective vaccine. Now it has been reduced to a few years and probably even less to develop a COVID-19 vaccine.

The genetic sequence of SARS-CoV-2 was published on January 11, 2020, triggering intense global research and development activity to develop a vaccine against the disease. The Coalition for Epidemic Preparedness Innovations (CEPI) is working with global health authorities and vaccine developers to support the development of vaccines against COVID-19.[2] Similarly, the WHO is facilitating[3] interactions between scientists, developers and funders to support coordination

DOI: 10.4324/9781003345091-9

and/or provide common platforms for working together. It has used its global mandate to rapidly convene 300 scientists, developers and funders to increase the likelihood that one or more safe and effective vaccines will soon be available to all. Numerous technologies are being utilized to develop a vaccine, including nucleic acid (DNA and RNA), virus-like particle, peptide, viral vector (replicating and nonreplicating), recombinant protein, live attenuated virus and inactivated virus approaches. According to the WHO, six vaccines are currently in clinical evalua-tion (with another about to start), and about 70 are in preclinical evaluation.[4]

Oxford University Vaccine Development

The Oxford University Vaccine Development programme reached clinical trials with its upcoming vaccine ChAdOx1 nCoV-19 in a short time. On April 23, 2020, it started testing this vaccine on human volunteers in Oxford, UK. Around 1110 people will take part in the trial, half receiving the vaccine and the other half (the control group) receiving a widely available meningitis vaccine.[5] ChAdOx1 nCoV-19 is based on an adenovirus vaccine vector and the SARS-CoV-2 spike protein, and it was produced in Oxford. It is made from a virus (ChAdOx1), which is a weakened form of the virus that causes the common cold (adenovirus) that causes infection in chimpanzees. However, it has been changed genetically so that it is not possible for the virus to grow in human beings. The study aims to test vac-cine safety and efficacy in healthy volunteers aged between 18 and 55. In a couple of months, the scientists are expecting to get the results.

India hopes to play a major role in manufacturing the vaccine in the near future. India is among the largest manufacturers of generic drugs and vaccines in the world. Serum Institute of India is the world's largest vaccine maker by number of doses produced and sold globally. It makes 1.5 billion doses every year, mainly from its two facilities in the western city of Pune. It has partnered to mass-produce this vaccine being developed by Oxford University. If all goes well, scientists hope to make at least a million doses by September 2020.[6]

Vaccine by CanSino Biologics

A recombinant novel COVID-19 vaccine based on an adenovirus vector 5 (Ad5-nCoV) has progressed fastest and has now entered phase 2 trials as of April 12, 2020. CanSino Biologics Inc. of China sponsored by the Institute of Biotechnol-ogy, Academy of Military Medical Sciences, PLA of China has raced ahead in the development of this vaccine.[7]

Ad5-nCoV is the most advanced genetically engineered DNA vaccine candi-date at the moment against COVID-19 and has been named as a top contender by the WHO. Phase 2 involves a randomized, double-blind and placebo-controlled clinical trial of healthy adults over 18 years of age. This clinical trial is designed to evaluate the immunogenicity and safety of Ad5-nCoV. The trial will enrol 500 subjects: 250 subjects in the middle-dose vaccine group, 125 in the low-dose group and 125 in the placebo group. Immunogenicity will be tested on days 0, 14 and 28 and again six months after vaccination.

Pfizer and BioNtech vaccine

On May 5, 2020, US pharmaceutical giant Pfizer, along with German biotech company BioNTech, began human trials of a potential mRNA COVID-19 vaccine (BNT 162) in the US.[8] The initial trial involved 360 volunteers, and the first subjects have already received injections. In addition to the US, some 200 patients have enrolled for trials in Germany.

The phase 1/2 study is designed to determine the safety, immunogenicity and optimal dose level of four mRNA vaccine candidates. The design of the trial will allow for the evaluation of the various mRNA candidates simultaneously, in order to identify the safest and potentially most efficacious candidate in a greater number of volunteers in the shortest possible amount of time:

> With our unique and robust clinical study program underway, starting in Europe and now the U.S., we look forward to advancing quickly and collaboratively with our partners at BioNTech and regulatory authorities to bring a safe and efficacious vaccine to the patients who need it most. The short, less than four-month timeframe in which we've been able to move from preclinical studies to human testing is extraordinary and further demonstrates our commitment to dedicating our best-in-class resources, from the lab to manufacturing and beyond, in the battle against COVID-19.
>
> *(Albert Bourla, chair and CEO, Pfizer)*[9]

During the trial, BioNTech will provide the clinical supply of the vaccine from its GMP-certified mRNA manufacturing facilities in Europe. If the vaccine development programme is successful, both companies plan to jointly produce and commercialize the vaccine. However, in China, BioNTech is already in collaboration with Shanghai Fosun Pharmaceutical (Group) Co., Ltd, to develop the same vaccine candidates. The companies are collaborating to conduct clinical trials in China, leveraging Fosun Pharma's extensive clinical development, regulatory and commercial capabilities in the country.

Vaccine by Moderna

Biotech company Moderna Inc. is trying to develop an mRNA vaccine against SARS-CoV-2. The vaccine candidate is named mRNA-1273. Around the middle of January 2020, the National Institute of Allergy and Infectious Diseases (NIAID), part of the National Institutes of Health (NIH), disclosed their intent to run a phase 1 study by using the mRNA-1273 vaccine. On March 16, 2020, the NIH announced that the first participant in its phase 1 study for mRNA-1273 was dosed, a total of 63 days from sequence selection to first human dosing. The phase

1 study was expected to provide information on immunogenicity and the safety of the vaccine. The open-label trial is expected to enrol 45 healthy adult volunteers aged 18 to 55 over approximately six weeks.[10]

In the middle of April 2020, the company further announced that the US Biomedical Advanced Research and Development Authority (BARDA) had agreed to donate up to $483 million to the company to accelerate the development of its mRNA vaccine candidate, mRNA-1273. On May 7, Moderna announced that it has been allowed to proceed to phase 2 study, and it is expected to begin soon. At the same time, it is finalizing the protocol for phase 3 study, which is expected to begin in early summer 2020.[11]

The vaccine development is progressing at a rapid pace, and a vaccine will probably be available by the end of 2020 or maybe in early 2021. However, there is a distinct possibility that the peak of the current pandemic will be over by that time. It will still be useful, though, because a second or third wave of the pandemic may occur. However, every vaccine development programme may not end in success. Vaccine failure is a grim reality of vaccine development programmes. If the vaccine is less than 60% successful in preventing infection, it may be considered a failure. Second, maintaining a cold chain is essential. If a vaccine is exposed to high temperature, during either transport or storage, it may become ineffective. Besides this, host factors like age, malnutrition and comorbid health conditions may make the vaccine ineffective.

The next important consideration in vaccine development is the ultimate cost of the vaccine. If the vaccine is priced beyond the reach of average people, then the ultimate purpose of the vaccine will not be fulfilled. A majority of the population in many developing and underdeveloped nation will not be able to afford the vaccine. At the same time, governments of many countries facing economic ruin due to the COVID-19 pandemic will not be in a position to vaccinate the entire population at its own cost. Thus, even after rapidly developing a vaccine against COVID-19, the threat of future pandemics will not be eliminated.

Given all these situations, it becomes clear that all the countries have to unite (preferably under the banner of the WHO) to develop, produce, distribute and immunize as many people of the world as possible.

Pandemic warning

The second most critical global challenge is pandemic warning. The vital questions that arise are were we forewarned about the current pandemic, and if yes, then why did we fail to pay heed to such a warning? Also, do we have a pandemic warning infrastructure in place already, or do we have to create one? The world has dealt with several pandemics in the past, including SARS, which was the first major pandemic in the 21st century. COVID-19 is neither the first nor will it be the last pandemic to affect humanity. US President Donald Trump was surprised that the United States was going to be badly affected by coronavirus. He was not ready to accept the fact and believed that the virus would go away in a few days. He said that it was unexpected.[12] However, the truth is far from that. Pandemic warnings have been issued at regular intervals in the recent past. Leading experts from the field of public health as well as industrialists like Bill Gates had given pandemic warnings[13] in the past and pointed out the under-preparedness of the American government to deal with a pandemic.[14]

Whenever there is a health crisis of a global nature, the US government and other international health agencies spring into action, and when the crisis resolves, urgency is replaced by complacency, as if nothing has happened and nothing is going to happen in the future. Funding for healthcare decreases; life goes on as usual; and gradually, everything fades from memory. People have forgotten the Spanish flu, SARS, MERS and swine flu. The most recent warning, a bipartisan report by the Center for Strategic and International Studies, was published on November 18, 2019. That report's prime recommendation was to undo the Trump administration's cuts to pandemic planning: 'Restore health security leadership at the White House National Security Council'. A careful review of past reports, books, planning documents, Congressional hearings, speeches and public testimonies makes clear that all the problems of the current COVID-19 crisis had been envisioned over the past 15 years.[15]

However, it is not the failure of the US government alone to pay heed to the dire warnings, but rather, it is a collective failure of world leaders, industrialists, health experts, policymakers and economists. Usually, the machinery for pandemic preparedness moves only when something happens; otherwise, it goes into a state of hibernation, only to be rudely jolted from sleep when the next pandemic strikes.

People have to face a deadly pandemic when those in positions of power choose to ignore the dire warnings. The pandemic strikes like a tsunami and kills or maims every susceptible person in its path.

GPMB annual report

The Global Preparedness Monitoring Board (GPMB), an independent monitoring and advocacy body, urges political action to prepare for and mitigate the effects of global health emergencies. The 15-member board is made up of political leaders, heads of agencies and experts, led jointly by Dr Gro Harlem Brundtland, former prime minister of Norway and director-general of the WHO, and Mr Elhadj As Sy, secretary general of the International Federation of Red Cross and Red Crescent Societies.

A dire warning about the pandemic threat was issued by GPMB in its September 2019 report:

> While disease has always been part of the human experience, a combination of global trends, including insecurity and extreme weather, has heightened the risk. Disease thrives in disorder and has taken advantage – outbreaks have been on the rise for the past several decades and the spectre of a global health emergency looms large. If it is true to say 'what's past is prologue', then there is a very real threat of a rapidly moving, highly lethal pandemic of a respiratory pathogen killing 50 to 80 million people and wiping out

nearly 5% of the world's economy. A global pandemic on that scale would be catastrophic, creating widespread havoc, instability and insecurity. The world is not prepared.[16]

Factors responsible for the pandemic

The various factors responsible for pandemic should be taken care of if we are to avoid a pandemic in the future. However, the main factor responsible for pandemic is us. We human beings are mainly responsible for pandemic. In this interconnected world, everyone is responsible for creating a pandemic-prone environment. The dangers are due to the following factors:[17]

- The world population is increasing towards 9.7 billion by 2050, and more and more people are expanding into wild frontiers.
- Unplanned urbanization and the rapid and mass movement of populations are forcing more people into overcrowded and unsanitary living conditions.
- Global warming allows disease vectors like mosquitoes and ticks to thrive over greater geographic areas.
- Insecurity and conflict are rising throughout the world, especially in places where outbreaks occur.

The population is growing continuously, and it is encroaching on the animal world, where animal pathogens come into contact with human beings and start circulating in them, leading to an outbreak. Thus, we are responsible for generating pandemics, and our close interaction and mobility help an outbreak to rapidly turn into a pandemic.

Steps for preventing and mitigating pandemics

The responsibility for preventing pandemics rests on everyone. It is a collective responsibility. Steps should be taken that are in line with the global public good, including people who work from local to international levels. The world needs to establish a system for early detection and control for this kind of threat. Reducing poverty should become a priority. Sustainable development goals need to be achieved by carefully using nonrenewable resources and maintaining an ecological balance. Environment degradation has to be avoided at all costs.

Leaders at all levels should be involved to devise a strategy for pandemic prevention. The governments of all countries should come under a global body like the UN or the WHO and must commit to and invest money into pandemic preparedness. Differences should be kept aside while doing so. Top global leaders should lead by example. Regional organizations should come forward and blaze a trail by setting examples. Governments should try to build a strong system involving all agencies and stakeholders, to prepare an effective mechanism for preparedness to deal with an outbreak.

Financial institutions should prepare policies to invest, in a regular manner, in enhancing as well as maintaining an effective mechanism for pandemic mitigation. The IMF and World Bank have to come forward to help developing countries in their efforts to build up their healthcare infrastructures. At the same time, these institutions should be prepared for the worst. Donors, global funds and philanthropists should come together and help the poor and vulnerable sections of society. Finally, the UN and the WHO should be at the forefront in fighting against pandemics.

The world paid the price for ignoring the pandemic warning. The infrastructure to forewarn against an impending pandemic is in place,[18] but global leaders failed to take these warnings seriously. As a matter of fact, the world has not been taking any steps to address the crucial issue of pandemic preparedness.

Public health

Public health is the science of protecting and improving the health of people and their communities. It is concerned with protecting the health of entire populations.[19] These populations can be as small as a local neighbourhood or as big as an entire country or region of the world. The next critical global challenge is the establishment and maintenance of an effective public health system for the world. The COVID-19 pandemic has refocused the attention of governments and policymakers on the role of universal healthcare through a public healthcare system all over the world. On May 6, 2020, the WHO in a media briefing said that the coronavirus crisis has revealed the importance of national and subnational health systems. WHO Director-General Dr Tedros Adhanom Ghebreyesus said that together these systems comprise 'the foundation of global health security'. He further said, 'Strong and resilient health systems are the best defence not only against outbreaks and pandemics, but also against the multiple health threats that people around the world face every day'.[20]

National public health services have played the major role in responding to the COVID-19 pandemic. They provide essential services to the government, to policymakers, to the healthcare system and to the general public. They spring into action the moment an outbreak occurs. With the help of experienced epidemiologists and a trained field force, the public health experts operate at the forefront to tackle the crisis. Door to door surveys, contact tracing, laboratory testing, epidemiological surveillance and analysis, providing information to government and helping policymakers, identifying vulnerable groups and providing information to the general public are some of the basic duties of public health experts. They also help the government to develop surge capacity in healthcare infrastructure, depending on the severity of the outbreak.

However, public health is in a state of decay due to decades of neglect by governments in various countries. This is because we do not fall ill together, but rather, we fall ill individually. Because of that, the world has moved towards the US system of insurance-based healthcare rather than government-backed public healthcare.

The COVID-19 pandemic made all of us sick at the same time, and in this collective fight, almost everybody has realized that private healthcare is not enough. Had it been enough, the morbidity and mortality would not have been so high in the US and certain European countries.[21]

Public health in developed countries

The US, the pioneer of private healthcare, has suffered the most from the COVID-19 pandemic. The death rate of Black Americans and Hispanic Americans is proportionately much higher than that of white Americans.[22] 'Slightly more than 70% of [coronavirus] deaths in Louisiana are African Americans', the state's governor, John Bel Edwards, said in a press conference. In Chicago, which is 30% Black, Black Americans account for 70% of all coronavirus cases in the city and more than half of the state's deaths. They were the most affected because they are the poorest, often work in essential services and do not enjoy the benefit of insurance-based healthcare. The COVID-19 pandemic has brought into focus the glaring fault of insurance-based private healthcare, and many Americans are now thinking about moving away from private healthcare to public healthcare.

In the UK, the government-run NHS is one of the best healthcare systems in the world. The majority of the British people were in favour of the NHS and were satisfied with its performance. However, many influential people and lawmakers were not so favourable towards it. Even Boris Johnson criticized the NHS in his book *Friends, Voters, Countrymen*. Besides this, the effective functioning of the NHS was hampered due to systemic defunding over several years.[23]

The NHS came to the rescue of Boris Johnson when he fell ill due to COVID-19, and the expert healthcare team of the NHS hospital cured his illness. After his recovery, Mr Johnson changed his mind and ultimately named his newborn child after the doctors who treated him during his hospital stay.[24] Whether this will further lead to a change in the future of the NHS is difficult to predict at this time.

Italy's Lombardy region had one of the best healthcare systems in the world, but the area turned out into one of the worst-hit regions. Lombardy changed its public health system to private healthcare in late 1990s.[25] Bergamo district led the way in this transformation. Unfortunately, it ended up with an unacceptable high mortality rate in Europe due to COVID-19. Hopefully, Italy has learnt a bitter lesson and will take appropriate steps in the near future.

Public health in India

The situation in India is not rosy at all. On one hand, India is credited with the successful eradication of smallpox and polio; on the other hand, it is criticized for its under-funding of its public health system. The Indian government's expenditure on health as a percentage of GDP still hovers around 1.5%, one of the lowest in the world.[26] India cannot be considered among the worst-hit countries given its low case fatality rate in comparison to its huge population. The existing public health system sprang into action early in response to the pandemic. However, the response was not uniform all over India. Containment strategies were devised, and contact tracing and field surveillance were carried out effectively to contain the outbreak. Infrastructure was ramped up, and hospital beds and ICU facilities were increased to take care of the surge in COVID-19 patients.

However, there were problems from the beginning as well. The people of India have little trust in government health facilities and the well-off have faith in the private healthcare system. At the same time, many feel that they are exploited by the greed of the private healthcare system. Many started running away from the quarantine, and others complained of the lack of facilities at the quarantine and COVID-19 care hospitals. The under-funded state-run healthcare system failed to rise up effectively but has still been trying to do a commendable job during this unprecedented crisis.

The good work done by the public health system of India seems to have been challenged by poverty. The government is in no position to lock up the entire nation and started relaxing lockdown measures for economic reasons. People started violating the lockdown rules, but the government was forced to look the other way. Within a few days of such relaxation, the number of cases started rising, and in the days to come, it has the potential to surpass even the US and turn the whole situation into a complete nightmare. The crucial issue that crept up in India was life versus livelihood versus the economy. The government of India had little choice and had to relax lockdown measures, and while easing lockdown, it remarked that people have been advised to learn to live with COVID-19.[27]

In spite of all the problems, the public health systems all over have tried to do their best to stop the deadly pandemic. The debate over whether a private health system or a public health system is better is going to increase in the coming days. Each one has its pros and cons. A strong public health system can go a long way towards saving lives during a pandemic, but it needs strong support from both the government and its citizens. Besides this, it can function only when the socioeconomic conditions of a country are good; otherwise, it will fail. It is now time for the governments of every country to refocus on public health, improve its functioning by providing proper funds and improve citizens' trust in the public health system. This global challenge needs to be addressed as soon as possible to prevent deaths and debility during this current pandemic as well as during future pandemics.

COVID-19 has now started badly affecting areas like Africa, Latin America and India. The narrow window of opportunity to contain this virus is practically over. Critical global challenges created by the COVID-19 pandemic have to be dealt with in a collective manner, with due urgency, to prevent further loss. Sharing data on the virus strains is crucial for developing an effective vaccine. The rapidly mutating virus will create problems in developing a vaccine which will be effective all over the world. Hence, collaboration rather than competition is the need of the hour.

Pandemics are global issues with global consequences. They need to be taken seriously by all concerned. Warning signs of a pandemic should not be swept under the rug. Warnings should be taken seriously, like any other challenging global issues. A pandemic has the capacity to kill or harm much more than even many wars or a natural disaster. Only global warming and pollution come anywhere near the destructive potential of a pandemic.

The development of a global public health system must now be the utmost priority. A pandemic poses a serious threat to global health security if large gaps in capacity and available resources continue to persist. This critical challenge of strengthening the public health system can be solved only by investing a substantial amount of money into public health. For this, an international financial institution like the IMF or the World Bank will have to play a crucial role. The rich nations have to help underdeveloped nations with aid. The WHO should in the near future come forward and advise all nations on how to devise a healthcare system to fight against pandemics. Without their help and global collaboration, it cannot be achieved.

Pandemics affect both the rich and poor and both developed countries and underdeveloped countries equally. The poor can pass the virus on to the rich and vice versa, or the virus may pass from an underdeveloped country to a developed country and vice versa. In such a situation, the rich or the developed nation should take the responsibility of pandemic preparedness and pass the benefit on to others. Global cooperation is the only solution to the global challenges emerging from the COVID-19 pandemic.

Notes

1 'Genetic recombinant technology' means the joining together of genetic material from different organisms and inserting it into a host organism to produce new *genetic* combinations that are of value to science, medicine, agriculture and industry
2 Le TT, Andreadakis Z, Kumar A, et al., The COVID-19 vaccine development landscape, *Nature*, available at www.nature.com/articles/d41573-020-00073-5, accessed on May 8, 2020
3 www.who.int/emergencies/diseases/novel-coronavirus-2019/global-research-on-novel-coronavirus-2019-ncov/solidarity-trial-accelerating-a-safe-and-effective-covid-19-vaccine WHO, accessed on May 8, 2020
4 Ibid.
5 Oxford COVID-19 vaccine begins human trial stage, available at www.ox.ac.uk/news/2020-04-23-oxford-covid-19-vaccine-begins-human-trial-stage, accessed on May 8, 2020
6 Biswas S, Coronavirus: How India will play a major role in a COVID-19 vaccine, *BBC*, available at www.bbc.com/news/world-asia-india-52363791, accessed on May 8, 2020
7 NIH, https://clinicaltrials.gov/ct2/show/NCT04341389, accessed on May 8, 2020
8 Pfizer and BioNTech begin human trial for COVID-19 vaccine in US, *Financial Express*, available at www.financialexpress.com/lifestyle/health/pfizer-and-biontech-begin-human-trial-for-covid-19-vaccine-in-us/1949331/, accessed on May 16, 2020
9 Global COVID-19 mRNA vaccine development program, *Business Wire*, available at www.businesswire.com/news/home/20200505005474/en/, accessed on May 8, 2020
10 Moderna, www.modernatx.com/modernas-work-potential-vaccine-against-covid-19, accessed on May 8, 2020

11 Ibid.

12 Graff GM, An oral history of the pandemic warnings Trump ignored, available at www.wired.com/story/an-oral-history-of-the-pandemic-warnings-trump-ignored/, accessed on May 8, 2020

13 Ibid.

14 Ibid.

15 Ibid.

16 GPMB annual report, available at https://apps.who.int/gpmb/assets/annual_report/GPMB_annualreport_2019.pdf, accessed on May 8, 2020

17 Ending the cycle of crisis and complacency in U.S. Global Health Security, available at https://healthsecurity.csis.org/final-report/, accessed on May 8, 2020

18 WHO continues to track the evolving infectious disease situation, sound the alarm when needed, share expertise, and mount the kind of response needed to protect populations from the consequences of epidemics, whatever and wherever might be their origin, available at www.who.int/csr/alertresponse/en/, accessed on May 16, 2020

19 What is public health? CDC Foundation, available at www.cdcfoundation.org/what-public-health, accessed on May 16, 2020

20 COVID-19 reveals gaps in health systems: WHO briefing, WHO, available at www.weforum.org/agenda/2020/05/covid-19-reveals-gaps-in-public-health-system-who-briefing/, accessed on May 8, 2020

21 Gupta D, Falling sick together: COVID-19 pandemic has immensely boosted the case for Universal Healthcare, *Times of India*, Editorial, April 30, 2020

22 The coronavirus can infect anybody but African Americans are dying in disproportionate numbers, especially in certain big cities, *The Guardian*, available at www.theguardian.com/world/2020/apr/08/its-a-racial-justice-issue-black-americans-are-dying-in-greater-numbers-from-covid-19, accessed on May 16, 2020

23 Government 'systematic underfunding' to blame for NHS humanitarian crisis, Labour says, *The Independent*, available at www.independent.co.uk/news/uk/politics/nhs-humanitarian-crisis-red-cross-labour-jeremy-hunt-underfunding-spending-cuts-a7514626.html, accessed on May 9, 2020

24 Boris Johnson and Carrie Symonds name baby son Wilfred Lawrie Nichola, *BBC*, available at www.bbc.com/news/uk-52513103, accessed on May 9, 2020

25 Pozzi G, Coronavirus puts into question Lombardy's long recognized healthcare efficiency, available at www.lavocedinewyork.com/en/news/2020/03/26/coronavirus-puts-into-question-lombardys-long-recognized-healthcare-efficiency/, accessed on May 9, 2020

26 Gap in India's preparedness for COVID-19 control, Patralekha Chatterjee, *Lancet*, available at www.thelancet.com/journals/laninf/article/PIIS1473-3099(20)30300-5/fulltext accessed on May 9, 2020

27 Dey S, Have to learn to live with COVID-19: Govt, *Times of India*, available at https://timesofindia.indiatimes.com/india/have-to-learn-to-live-with-covid-19-govt/articleshow/75638429.cms, accessed on May 9, 2020

10

OLD AND NEW CORONAVIRUS VARIANTS

Alpha, Beta, Gamma, Delta and Omicron

The COVID-19 pandemic from its origin in Wuhan, China, spread rapidly all over the globe in a short span of time and emerged as the most catastrophic health crisis since the influenza pandemic of 1918. The initial spread, epidemiology, clinical manifestations, treatment and efforts of vaccine development have been written in the previous chapters. However, a lot of new developments, especially the emergence of variants of SARS-CoV-2, have occurred, which need to be understood for leading a life in presence of the ongoing pandemic which is not yet showing signs of ebbing all over the world.

All viruses change with time. SARS-CoV-2 has also changed several times during its advance throughout the world over the last two years. During replication[1] of the RNA genome, some changes are introduced sporadically that alter its genome and result in a subpopulation of defective virus genome that may or may not result in evolution and stability within a host. When these mutations[2] are well established and successfully fixed in a viral population within a host and successfully transmitted in other individuals, they start competing with the original virus population already in circulation.[3] These new crop of variants may affect disease severity, transmissibility, diagnostic tests, vaccines and treatment.[4] In spite of substantial progress in understanding the mechanism of the disease-causing process by SARS-CoV-2 and improvement in management of COVID-19, the variants are causing havoc all over the globe leading to several outbreaks in the last two years. These variants to a great extent have blunted the significant efforts made to halt the global spread of the pandemic.

These variants have been emerging and circulating throughout the globe since the beginning of the pandemic. The World Health Organization (WHO), in collaboration with partners, expert networks, national authorities, institutions and researchers have been monitoring and assessing the evolution of SARS-CoV-2 since January 2020.[5] The WHO is doing this to inform the countries and the public in

DOI: 10.4324/9781003345091-10

advance to prevent the global spread. These have been broadly classified as either variants under monitoring, variants of interest or variants of concern.

In the UK, variants are only classified as variants of concern by the UK Health Security Agency (UKHSA). In the US, variants are classified by the Centers for Disease Control and Prevention (CDC) as variants being monitored, variants of interest, variants of concern or variants of high consequence.[6] The WHO has assigned the Greek alphabet for key variants like Alpha, Beta, Gamma, Delta and Omicron. This however, does not replace existing scientific names (e.g., Pango, Nextstrain, GISAID), which continue to be used in research. The genomic sequence of original SARS-CoV-2 identified from the first case (December 2019) is termed 'index virus'.[7]

Variant of interest

WHO has defined a SARS-CoV-2 variant of interest (VOI) as a variant:

- with genetic changes that are predicted or known to affect virus characteristics such as transmissibility, disease severity, immune escape, or diagnostic or therapeutic escape; and
- identified to cause significant community transmission or multiple case clusters, in multiple countries with increasing relative prevalence alongside increasing number of cases over time, or other apparent epidemiologic impacts to suggest an emerging risk to global public health.[8]

At present there are no variants of interest in circulation. Previously circulating variants of interest include the Epsilon, Zeta, Eta, Theta, Iota, Kappa, Lambda and Mu variants. If a variant of interest is detected, the WHO will do a comparative assessment of variant characteristics and public health risks. If determined necessary, coordinated laboratory investigations will be done with Member States and partners. There will be a review of global epidemiology of a variant of interest, and the WHO will monitor and track its global spread.[9]

Variant of concern

WHO has provided a working definition of variant of concern (VOC). It has defined that a SARS-CoV-2 variant meets the definition of a variant of interest and through a comparative assessment, has been demonstrated to be associated with one or more of the following changes at a degree of global public health significance:[10]

- Increase in transmissibility or detrimental change in COVID-19 epidemiology; or
- Increase in virulence or change in clinical disease presentation; or
- Decrease in effectiveness of public health and social measures, or available diagnostics, vaccines, therapeutics.

At present the variants of concern according to the WHO and the CDC are the Delta and Omicron variants. Omicron variants include BA.1, BA.2, BA.3, BA.4, BA.5 and descendent lineages. It also includes BA.1/BA.2 circulating recombinant forms such as XE. Previously circulating variants of concern include the Alpha, Beta and Gamma variants.

On detection of a potential variant of concern by the WHO and Member States, the primary actions taken are the following:

- Comparative assessment of variant characteristics and public health risks by the WHO and the Technical Advisory Group on Virus Evolution;
- If determined necessary, coordination of additional laboratory investigations with Member States and partners;
- Communication of new designations and findings with Member States and the public through established mechanisms;
- Evaluation of WHO guidance through established WHO mechanisms and updates, if necessary.

Alpha variant

The Alpha variant was previously nominated by the WHO as a variant of concern. It is also known as B.1.1.7 according to Pango lineage.[11] It was initially detected in September 2020 in Kent, South-East of England. It was nominated as a variant of concern by the WHO on December 18, 2020.[12]

Alpha was initially able to spread more rapidly[13] than other variants locally present across the globe and toward the end of 2020 it replaced the other circulating strains and became dominant in most of Europe.[14] The Alpha variant includes 17 mutations in the viral genome. Of these, eight mutations are in the spike protein.[15] The transmissibility of Alpha variant was estimated to be 1.5 times higher than that of other circulating lineages in the UK at that time. It had enhanced 35 times more affinity to bind to ACE2 receptors.[16] By March 2021, it became the predominant variant in the US.[17] The WHO reported the presence of Alpha variant in 130 countries by March 30, 2021.[18] By mid-August 2021 it had spread to more than 185 countries.[19]

The severity of disease caused by Alpha-variant has conflicting data,[20] although most authors agree that this VOC is associated with increased hospitalization rates.[21] High intensive care admission rate indicates that the disease seems to be more severe, but not mortality as compared to the wild virus.[22] However, a study published in nature has shown that it has higher mortality rate[23] as compared to the other SARS-CoV-2 infection.[24] The symptoms and signs of patients suffering from COVID-19 due to Alpha variant were more or less similar to that caused by the previously circulating virus.[25] One study found that the AstraZeneca vaccine was 70% effective in preventing symptomatic COVID-19 caused by the Alpha variant, while another study estimated the efficacy of the Pfizer vaccine at roughly 90%.[26]

Beta variant

The Beta variant was previously classified as a variant of concern by the WHO. It is also known as B.1.351 in Pango lineage. It has eight mutations in its spike protein. This lineage was identified in South Africa after the first wave of the epidemic in a severely affected metropolitan area (Nelson Mandela Bay) that is located on the coast of the Eastern Cape province, South Africa in October 2020. This lineage spread rapidly, and became dominant in the Eastern Cape, Western Cape and KwaZulu-Natal provinces within weeks.[27] The earliest documented sample was in early August 2020.[28] By the end of January 2021 it reached the US. The WHO reported the presence of the Beta variant in 80 countries by March 30, 2021.[29]

The Beta variant is no more transmissible than the Alpha variant.[30] However, it was linked with a rise in hospitalizations and deaths during South Africa's second wave. Vaccines also appear to be less effective in preventing COVID-19 from the Beta variant. In one study, researchers found that two doses of the Pfizer vaccine were 75% effective against any infection from the Beta variant. However, vaccine effectiveness against severe or fatal disease from the variants was very high at 97.4%.[31]

Gamma variant

WHO had previously classified the Gamma variant as a variant of concern.[32] It is known as P.1 in Pango lineage. The earliest sample was documented in Brazil/ Japan in December 2020.[33] It was estimated to be 1.7–2.4 times more transmissible than other local strains in Brazil. The Gamma variant has some of the same mutations in its spike protein as the Alpha and Beta strains, which allow it to attach more easily to human cells.[34] Studies indicate that this new variant has greater transmission capacity, even though it is not associated with more severe clinical conditions in relation to the original strain.[35] The WHO reported the presence of the Gamma variant in 45 countries by March 30, 2021.[36]

Delta variant

The Delta variant is classified as a variant of concern by the WHO, CDC and UKHSA.[37] Pango lineage of the Delta variant is B.1.617.2 (including all AY sublineages). The earliest documented sample was detected in India in October 2020. It was initially classified as a variant of interest by the WHO but later it was classified as variant of concern in May 2021 when it spread rapidly and caused the brutal second wave in India and many other countries throughout the world.[38] In the United States, this variant was first detected in March 2021. By mid-2021, it had been detected in more than 142 countries. The B.1.617.2 variant has ten mutations in the spike protein;[39] however, mutation P681R located at a furin cleavage site leads to increased infection via cell surface entry.[40] Studies out of the United Kingdom suggest the Delta variant is up to 60% more transmissible than the Alpha variant.

In the UK, the secondary attack rate among household contacts of cases that have not travelled was 11.3%, compared with 10.2% with Alpha (as of November 22, 2021).[41]

Besides increased transmissibility, the Delta variant has a shortened incubation time, enhanced replication abilities and is more infectious compared to non-VOCs.[42] It has been reported that viral load is 1260 times higher in Delta variant patients than the original COVID-19 strain.[43] After its emergence it became the most important variant and was mainly responsible for the huge spike in the number of cases across the globe in the middle of 2021. In India, the Delta-driven wave started in March 2021 and ebbed by early July, leaving behind thousands of deaths. In the UK, where the vaccination rate was as high as 76% in adults, the Delta-fuelled wave that started in June began to recede by August 2021. Similarly, in the US where 50% of the population was fully vaccinated, by late July about an 80–87% increase in cases (seven-day average of 92,000 cases daily) was caused by the Delta variant.[44] A study published in the famous medical journal *Lancet* has shown that the household secondary attack rate among unvaccinated contacts exposed to index cases infected with the Delta variant was higher at 25.8% compared with 12.9% among unvaccinated contacts exposed to other variants. There is a significantly higher risk of infection among unvaccinated close contacts by the Delta variant compared to other variants, which would explain the rapid and extensive epidemic in India.[45]

A study published in the CDC's Morbidity and Mortality Weekly Report, reported that fully vaccinated people accounted for 74% of nearly 469 COVID-19 cases (4/5 subjects hospitalized were fully vaccinated, and no one died).[46] Importantly, this study found that fully vaccinated people carried the same amount of Delta variant virus in their noses/throats as the unvaccinated.[47]

The symptoms and signs were similar to the general set of symptoms of COVID-19 but the disease was much more severe in the unvaccinated as compared to the vaccinated.[48] The observation that the Delta variant can infect vaccinees (therefore, can spread the infection) led the CDC to change their recommendation that even fully vaccinated people should wear masks in indoor public places. Studies in Korea, Spain and Germany have found that SARS-CoV-2 transmission was more common from adults to children than from children to adults. Among the contacts there is a lower risk of being infected by SARS-CoV-2 for fully vaccinated compared to unvaccinated contacts, and a lower risk of symptomatic disease in both fully and partially vaccinated contacts.[49]

The Pfizer vaccine has 64% efficacy against infection and preventing symptomatic cases, but is 93% effective at preventing hospitalization/severe disease caused by the Delta variant. As of August 2021, another set of data showed that efficacy of Moderna vaccine dropped from 86% to 76% against infection and Pfizer had dropped to 42%, while the Delta variant was evolving as the dominant strain.[50]

Another study published in the reputed journal *NEJM* showed that the effectiveness was approximately 88% with two doses of the BNT162b2 vaccine and approximately 67% with two doses of the ChAdOx1 nCoV-19 (Oxford/Astra

Zeneca) vaccine against the Delta variant. A clear effect was noted with both vaccines, with high levels of effectiveness after two doses.[51]

According to *BMJ Best Practice*, the disease severity appears to be associated with an increased risk of hospitalization (suggesting more severe disease) compared with contemporaneous Alpha cases; however, there is a high level of uncertainty in these findings.[52] There was a high prevalence of pneumonia[53] in patients affected by the Delta variant and in majority of the cases, the disease turned out to be severe with a fall in oxygen saturation requiring oxygen support and ventilation. The crude case fatality rate is estimated to be 0.53%, considerably less than the Alpha variant (as of October 26, 2021).[54] CDC data published in October 2021 states that the proportion of non-pregnant adults aged ≥18 years hospitalized with COVID-19 who were admitted to an ICU or died during their hospitalization did not significantly change during this period. No significant differences in severity were observed between the pre-Delta and Delta periods among fully vaccinated or unvaccinated hospitalized patients, overall or when stratified by age and vaccination status. However, during the Delta period, adults aged 18–49 years accounted for a larger proportion of hospitalized patients compared with the pre-Delta period. This was driven by the larger number of unvaccinated hospitalized patients in this age group, likely reflecting lower vaccination coverage in younger adults than in older adults.[55]

Even if data from the Western World reflect that death due to the Delta variant was not substantial, the grim reality seems to portray a different picture altogether from India which faced the brutal onslaught of the Delta variant in April–May 2021. A study published in the reputed medical journal *Lancet* estimated that 5.2 excess deaths occurred per 1000 residents during the COVID-19 pandemic in Chennai, a mega city in South India, representing a 41% increase over typical mortality levels. Most excess deaths occurred during the second wave of the pandemic, when mortality peaked at levels 4.75 times higher than pre-pandemic observations.[56] There is a disagreement between the Indian government and the WHO regarding the actual death in India during the ongoing pandemic. On May 5, 2022, the WHO released its official data on excess mortality and reported that there were likely 4.7 million deaths directly or indirectly attributable to COVID-19 in India in 2020 and 2021, which is almost ten times the official government COVID-19 death count. India has objected to this data saying the methodology used was not correct.[57] Until the real picture comes out it will be premature to conclude that the Delta variant mortality was not high compared to other variants or was at par with them.

Omicron variant

The Omicron variant (Pango lineage B.1.1.529), first detected in South Africa in November 2021,[58] is currently classified as a variant of concern by the WHO, CDC and the UKHSA.[59] On November 24, 2021, South Africa reported the identification of the Omicron variant to the WHO. It was first detected in specimens collected on November 11, 2021, in Botswana and on November 14, 2021, in

South Africa.[60] However, it is probable that there were unidentified cases in several countries across the world before then.

Omicron was quickly designated as a VOC on November 26, 2021, by the WHO due to more than 30 changes to the spike protein of the virus, along with the sharp rise in the number of cases observed in South Africa. On December 1, 2021, the first case attributed to the Omicron variant was reported in the United States in a person who returned from travel to South Africa. The first Omicron case in England was reported on November 27, 2021.[61] Since then the Omicron variant has become the dominant variant in many countries. To date, this variant has been identified in 133 countries and is now the most prevalent lineage globally, representing 85% of variant cases reported in late January 2022.[62]

Omicron has a substantial growth advantage over Delta, and has rapidly replaced Delta globally. There is significant evidence that immune evasion contributes to its rapid spread, but it is unknown how much intrinsic increased transmissibility contributes and further research is required.[63] An artificial intelligence model has revealed that Omicron may be over 10 times more contagious than the original virus or about 2.8 times as infectious as the Delta variant.[64] Omicron is characterized by several mutations of the spike protein. The small body of evidence available suggests that these mutations result in increased transmissibility when compared with the wild type and previous VOCs, and reduced potency of neutralizing antibodies.[65] According to the CDC the Omicron variant spreads more easily than earlier variants of the virus that cause COVID-19, including the Delta variant. The CDC expects that anyone with an Omicron infection, regardless of vaccination status or whether or not they have symptoms, can spread the virus to others.[66]

Data on clinical severity of patients infected with Omicron is growing, but is still limited. Early data from South Africa, the UK, Canada and Denmark suggested a reduced risk of hospitalization for Omicron compared with Delta.[67] Data from the US also supports this trend, but acknowledges that a higher number of cases (5 times higher than the Delta wave) due to increased transmissibility of the variant is resulting in a record number of hospitalizations (1.8 times higher compared with the Delta wave).[68]

One of the largest and latest studies published in the *Lancet* has reported that confirmed Omicron cases had a 59% lower risk of hospital admission, a 44% lower risk of any hospital attendance, and a 69% lower risk of death than that of confirmed Delta cases. In those over 20 years of age, there was a significant reduction in the risk of hospitalization for Omicron compared with Delta. In cases aged 50 years and above, the estimated reduction in the risk of hospitalization was found to be 50–75%. However, the magnitude of severity reduction was found to be lower for those older than 80 years, but still over 50% for the various end points studied. Besides this it was further published that in children aged 0–9 years with a confirmed infection, the risk of disease sufficiently severe enough to result in death was very low, and it was estimated that the risk of hospitalization from Omicron infection was not significantly different from that of Delta infection.[69] Laboratory studies have shown that Omicron replicates more in upper airway cells and less in the lungs which is the most probable cause of less severity of disease other than widespread vaccination and previous infection in many countries.[70]

It was further observed that the proportion of hospitalized COVID-19 patients requiring intensive care or mechanical ventilation (or both) has been substantially lower during the Omicron wave in England than the preceding Delta wave. Overall, 80% reduction in the intrinsic risk of death was estimated for Omicron infection compared with that of Delta.[71]

Early data suggests that the vaccine efficacy is significantly lower against Omicron infection and symptomatic disease compared with Delta, with homologous and heterologous booster doses increasing vaccine effectiveness. After two doses, vaccine effectiveness waned rapidly, with very limited vaccine effects seen from 20 weeks after the second dose of any vaccine.[72] Vaccine efficacy estimates against severe outcomes (e.g., hospitalization) are lower for Omicron compared with Delta, but mostly remain >50% after the primary series and improve with a booster dose to >80%.[73] However, it is uncertain how long this increased protection lasts for. Increased risk of re-infection has been reported by South Africa, the UK, Denmark and Israel.[74] In general vaccine effectiveness was lower for the Omicron variant than for the Delta variant at all intervals after vaccination and for all combinations of primary courses and booster doses investigated.[75]

Omicron sublineage

The Omicron variant comprises five known lineages including the parental lineage B.1.1.529, and the descendent sublineages BA.1, BA.1.1, BA.2 and BA.3.[76] The WHO and its partners are monitoring all these lineages. Of them, the most common ones are BA.1, BA.1.1 and BA.2. The WHO's Technical Advisory Group on SARS-CoV-2 Virus Evolution (TAG-VE) has reinforced that BA.2 sublineage should continue to be considered as a VOC and that it should remain classified as Omicron.[77]

BA.2 differs from BA.1 in its genetic sequence; BA.2 has fewer mutations than BA.1, with 31 on the spike protein and other proteins.[78] Studies have shown that BA.2 has a growth advantage over BA.1. Studies are ongoing to understand the reasons for this growth advantage, but initial data suggest that BA.2 appears inherently more transmissible than BA.1, which currently remains the most common Omicron sublineage reported. The WHO reported on February 1, 2022, that BA.2 designated sequences have been detected from 57 countries to date, with the weekly proportion of BA.2 relative to other Omicron sequences rising to over 50% during the last six weeks in several countries.[79] Further, although BA.2 sequences are increasing in proportion relative to other Omicron sublineages (BA.1 and BA.1.1), there is still a reported decline in overall cases globally. Reinfection with BA.2 following infection with BA.1 has been documented; however, initial data from population-level reinfection studies suggest that infection with BA.1 provides strong protection against reinfection with BA.2.[80]

BA.2 has demonstrated an increased growth rate compared with BA.1.[81] However, there is no reported difference in severity or hospitalization between BA.2 and BA.1.[82] Similarly, there is no reported difference in vaccine efficacy between BA.2 and BA.1.[83] At least 25 weeks after two doses, vaccine effectiveness against symptomatic infection was reported as 13% for BA.2 (versus 9% for BA.1). At two weeks after a third booster dose this increased to 70% (versus 63% for BA.1).[84]

Not much is known about BA.3 lineage. Less than 400 cases have been reported as on February 3, 2022, in 19 countries. BA.3 has 33 mutations in its spike protein. Not much is till known about disease severity and vaccine effectiveness.[85]

South Africa is seeing a wave of COVD-19 infections linked to two new forms, or 'sub-lineages', of Omicron: BA.4 and BA.5. BA.4 and BA.5 have now been found in Europe and the UK,[86] Australia,[87] USA and Botswana.[88]

The Omicron lineage BA.4 was first detected from a specimen collected on January 10, 2022, in Limpopo, South Africa and by April 29, 2022, BA.4 had been detected in all the provinces of the country. The percentage of sequences designated BA.4 has grown from <1% in January 2022 to >35% in April 2022.[89]

The Omicron lineage BA.5 was first detected from a specimen collected on February 25, 2022, in KwaZulu-Natal. By April 22, 2022, BA.5 had been detected in Gauteng, Limpopo, Mpumalanga, KwaZulu-Natal, the North West, and the Western Cape in South Africa. The percentage of sequences designated BA.5 has increased from <1% in January 2022 to 20% in April 2022.[90] Collectively, the percentage of sequences assigned to BA.4 and BA.5 have increased from <1% in January 2022 to >50% in April 2022. As on May 7, 2022, the proportion of positive new cases/total new tested was (31.1%). The seven-day moving average daily number of cases has increased as compared to the past.[91] This trend clearly points towards a high likelihood of these two Omicron subvariants becoming the dominant strain globally in the near future.

Recombinant variants

A recombinant virus is a combination of two previously existing virus strains. In the case of the SARS-CoV-2 virus, the recombinant strain shares the genetic material of two strains, which in the case of currently circulating hybrid variants is Delta and BA.1 variants, or BA.1 and BA.2 variant. It is a natural phenomenon that occurs when two viruses remain in the same cell at the same time, allowing the variants to interact during replication, mix up their genetic material, form new combinations which can be considered a mutational event. Two Delta and Omicron recombinants and one BA.1 × BA.2 recombinant have now been given Pango lineage designations XD, XF and XE. Of these, XD and XF are recombinants of Delta and Omicron BA.1, while XE is a recombinant of Omicron BA.1 and BA.2. The WHO continues to monitor these recombinant variants. None of the preliminary available evidence indicates that these recombinant variants are associated with higher transmissibility or more severe outcomes.[92]

Variants will continue to emerge in future. Some variants will emerge and disappear while others will persist. Even if a variant causes less severe disease in general, an increase in the overall number of cases could cause an increase in hospitalizations, put more strain on healthcare resources and potentially lead to more deaths.[93] Hence, global efforts to detect, track and monitor the emerging variants along with development of effective treatment and better vaccines in the future are absolutely necessary to put a brake on the ongoing pandemic.

Notes

1 *Replication* is the process of formation of new *viruses* during the infection process in the target host cells.
2 A change in the structure of the genes or chromosomes of an organism.
3 Wassenaar TM, et al., The first three waves of the Covid-19 pandemic hint at a limited genetic repertoire for SARS-CoV-2, available at https://academic.oup.com/femsre/advance-article/doi/10.1093/femsre/fuac003/6514530?login=false, accessed on March 6, 2022.
4 Coronavirus Disease 2019 (COVID-19), *BMJ Best Practice*, available at https://bestpractice.bmj.com/topics/en-us/3000168/aetiology, accessed on April 20, 2022.
5 Tracking SARS-CoV-2 variants, WHO, available at www.who.int/en/activities/tracking-SARS-CoV-2-variants/, accessed on April 20, 2022.
6 SARS-CoV-2 variant classifications and definitions, CDC, available at www.cdc.gov/coronavirus/2019-ncov/variants/variant-classifications.html, accessed on April 20, 2022.
7 Tracking SARS-CoV-2 variants, WHO, available at www.who.int/en/activities/tracking-SARS-CoV-2-variants/, accessed on April 20, 2022.
8 Ibid.
9 Ibid.
10 Ibid.
11 Pango lineage names comprise an alphabetical prefix and a numerical suffix.
12 Tracking SARS-CoV-2 variants, WHO, available at www.who.int/en/activities/tracking-SARS-CoV-2-variants/, accessed on April 20, 2022.
13 Aleem A, et al., Emerging variants of SARS-CoV-2 and novel therapeutics against coronavirus (COVID-19), available at https://pubmed.ncbi.nlm.nih.gov/34033342/, accessed on April 21, 2022.
14 Davies NG, et al., Estimated transmissibility and impact of SARS-CoV-2 lineage B.1.1.7 in England, *Science*, 2021; 372(6538): eabg3055, available at https://pubmed.ncbi.nlm.nih.gov/33658326/, accessed on April 20, 2022.
15 Cascella M, et al., Features, evaluation, and treatment of coronavirus (COVID-19), available at www.ncbi.nlm.nih.gov/books/NBK554776/, accessed on April 27, 2022.
16 Wassenaar TM, et al., The first three waves of the Covid-19 pandemic hint at a limited genetic repertoire for SARS-CoV-2, available at https://academic.oup.com/femsre/advance-article/doi/10.1093/femsre/fuac003/6514530?login=false, accessed on March 6, 2022.
17 Galloway SE, et al., Emergence of SARS-CoV-2 B.1.1.7 lineage – United States, December 29, 2020–January 12, 2021, available at https://pubmed.ncbi.nlm.nih.gov/33476315/, accessed on April 20, 2022.
18 COVID-19 weekly epidemiological update, WHO, available at https://www.who.int/docs/default-source/coronaviruse/situation-reports/20210330_weekly_epi_update_33.pdf?sfvrsn=f2c2d63d_16&download=true, accessed on April 22, 2022.
19 COVID-19 weekly epidemiological update, WHO, available at https://www.who.int/docs/default-source/coronaviruse/situation-reports/20210810_weekly_epi_update_52.pdf?sfvrsn=8ae11f92_3&download=true, accessed on April 22, 2022.

20 England, Grint DJ, et al., Severity of severe acute respiratory system coronavirus 2 (SARS-CoV-2) alpha variant (B.1.1.7), *Clinical Infectious Disease*, available at https://academic.oup.com/cid/advance-article/doi/10.1093/cid/ciab754/6365124, accessed on April 20, 2022.

21 Domingo P, Alpha variant SARS-CoV-2 infection: How it all starts, *Lancet*, available at www.thelancet.com/journals/ebiom/article/PIIS2352-3964(21)00497-7/fulltext, accessed on April 20, 2022.

22 Coronavirus Disease 2019 (COVID-19), *BMJ Best Practice*, available at https://bestpractice.bmj.com/topics/en-us/3000168/aetiology, accessed on April 20, 2022.

23 Davies NG, et al., Increased mortality in community-tested cases of SARS-CoV-2 lineage B.1.1.7, *Nature*, 2021; 593(7858): 270–274.

24 Ibid.

25 Coronavirus Disease 2019 (COVID-19), *BMJ Best Practice*, available at https://bestpractice.bmj.com/topics/en-us/3000168/aetiology, accessed on April 20, 2022.

26 Duong D, Alpha, Beta, Delta, Gamma: What's important to know about SARS-CoV-2 variants of concern? *CMAJ*, 2021; 193(27): E1059–E1060, available at www.ncbi.nlm.nih.gov/pmc/articles/PMC8342008/, accessed on April 22, 2022.

27 Tegally H, et al., Detection of a SARS-CoV-2 variant of concern in South Africa, available at https://pubmed.ncbi.nlm.nih.gov/33690265/, accessed on April 22, 2022.

28 COVID-19 weekly epidemiological update, WHO, available at https://www.who.int/docs/default-source/coronaviruse/situation-reports/20210330_weekly_epi_update_33.pdf?sfvrsn=f2c2d63d_16&download=true, accessed on April 22, 2022.

29 Ibid.

30 Duong D, Alpha, Beta, Delta, Gamma: What's important to know about SARS-CoV-2 variants of concern? *CMAJ*, 2021; 193(27): E1059–E1060, available at www.ncbi.nlm.nih.gov/pmc/articles/PMC8342008/, accessed on April 22, 2022.

31 Ibid.

32 Tracking SARS-CoV-2 variants, WHO, available at www.who.int/en/activities/tracking-SARS-CoV-2-variants/, accessed on April 20, 2022.

33 COVID-19 weekly epidemiological update, WHO, available at https://www.who.int/docs/default-source/coronaviruse/situation-reports/20210330_weekly_epi_update_33.pdf?sfvrsn=f2c2d63d_16&download=true, accessed on April 22, 2022.

34 Duong D, Alpha, Beta, Delta, Gamma: What's important to know about SARS-CoV-2 variants of concern? *CMAJ*, 2021; 193(27): E1059–E1060, available at www.ncbi.nlm.nih.gov/pmc/articles/PMC8342008/, accessed on April 22, 2022.

35 Claudio da Silva J, New Brazilian variant of the SARS-CoV-2 (P1/Gamma) of COVID-19 in Alagoas state, *Br J Inf Dis*, May–June 2021, available at www.bjid.org.br/en-new-brazilian-variant-sars-cov-2-p1-gamma-articulo-S141386702100057X, accessed on April 22, 2022.

36 COVID-19 weekly epidemiological update, WHO, available at https://www.who.int/docs/default-source/coronaviruse/situation-reports/20210330_weekly_epi_update_33.pdf?sfvrsn=f2c2d63d_16&download=true, accessed on April 22, 2022.

37 Coronavirus Disease 2019 (COVID-19), *BMJ Best Practice*, available at https://bestpractice.bmj.com/topics/en-us/3000168/aetiology, accessed on April 20, 2022.

38 Aleem A, et al., Emerging variants of SARS-CoV-2 and novel therapeutics against coronavirus (COVID-19), available at https://pubmed.ncbi.nlm.nih.gov/34033342/, accessed on April 21, 2022.

39 Cascella M, et al., Features, evaluation, and treatment of coronavirus (COVID-19), available at www.ncbi.nlm.nih.gov/books/NBK554776/, accessed on April 27, 2022.

40 Liu Y, et al., Delta spike P681R mutation enhances SARS-CoV-2 fitness over Alpha variant, available at https://pubmed.ncbi.nlm.nih.gov/34462752/, accessed on April 25, 2022.

41 Coronavirus Disease 2019 (COVID-19), *BMJ Best Practice*, available at https://bestpractice.bmj.com/topics/en-us/3000168/aetiology, accessed on April 20, 2022.

42 Reardon S, How the Delta variant achieves its ultrafast spread, *Nature* 2021 [cited August 11, 2021]. https://doi.org/10.1038/d41586-021-01986-w.

43 Ibid.
44 Yu F, et al., COVID-19 Delta variants – Current status and implications as of August 2021, *Prec Clin Med*, 2021; 4(4): 287–292, https://doi.org/10.1093/pcmedi/pbab024, available at, https://academic.oup.com/pcm/article/4/4/287/6372920?login=false, accessed on March 6, 2022.
45 Ng OT, et al., Impact of Delta variant and vaccination on SARS-CoV-2 secondary attack rate among household close contacts, *Lancet*, November 1, 2021, available at www.thelancet.com/journals/lanwpc/article/PIIS2666-6065(21)00208-X/fulltext, accessed on April 22, 2022.
46 Brown CM, et al., Outbreak of SARS-CoV-2 infections, including COVID-19 vaccine breakthrough infections, associated with large public gatherings – Barnstable County, Massachusetts, July 2021, *MMWR Morb Mortal Wkly Rep*, 2021; 70: 1059–1062, https://doi.org/10.15585/mmwr.mm7031e2, available at www.cdc.gov/mmwr/volumes/70/wr/mm7031e2.htm?s_cid=mm7031e2_w#suggestedcitation, accessed on April 22, 2022.
47 Ibid.
48 Ibid.
49 Ng OT, et al., Impact of Delta variant and vaccination on SARS-CoV-2 secondary attack rate among household close contacts, *Lancet*, November 1, 2021, available at www.thelancet.com/journals/lanwpc/article/PIIS2666-6065(21)00208-X/fulltext, accessed on April 22, 2022.
50 Yu F, et al., COVID-19 Delta variants – Current status and implications as of August 2021, *Prec Clin Med*, available at https://academic.oup.com/pcm/article/4/4/287/6372920?login=false, accessed on April 23, 2022.
51 Lopez Bernal J, et al., Effectiveness of Covid-19 vaccines against the B.1.617.2 (Delta) variant, *NEJM*, available at www.nejm.org/doi/full/10.1056/nejmoa2108891, accessed on April 23, 2022.
52 Coronavirus Disease 2019 (COVID-19), *BMJ Best Practice*, available at https://bestpractice.bmj.com/topics/en-us/3000168/aetiology, accessed on April 20, 2022.
53 Hu K, et al., COVID-19: Risk factors for severe cases of the Delta variant, available at www.ncbi.nlm.nih.gov/pmc/articles/PMC8580340/, accessed on April 23, 2022.
54 Ibid.
55 Severity of Disease among Adults Hospitalized with Laboratory-Confirmed COVID-19 before and during the Period of SARS-CoV-2 B.1.617.2 (Delta) Predominance – COVID-NET, 14 States, January–August 2021, *MMWR*, CDC, available at www.cdc.gov/mmwr/volumes/70/wr/mm7043e1.htm, accessed on April 23, 2022.
56 Lewnard JA, et al., All cause mortality during the COVID-19 pandemic in Chennai, India: A observational study, *Lancet*, available at www.thelancet.com/journals/laninf/article/PIIS1473-3099(21)00746-5/fulltext, accessed on April 23, 2022.
57 Koshy J, WHO estimates 4.7 million COVID-19-linked deaths in India, *The Hindu*, available at www.thehindu.com/sci-tech/health/who-estimates-47-million-covid-linked-deaths-in-india-10-times-official-count/article65385669.ece, accessed on May 6, 2022.
58 Aleem A, et al., Emerging variants of SARS-CoV-2 and novel therapeutics against coronavirus (COVID-19), available at https://pubmed.ncbi.nlm.nih.gov/34033342/, accessed on April 21, 2022.
59 Coronavirus Disease 2019 (COVID-19), *BMJ Best Practice*, available at https://bestpractice.bmj.com/topics/en-us/3000168/aetiology, accessed on April 20, 2022.
60 Science brief: Omicron (B.1.1.529) variant, CDC, available at www.cdc.gov/coronavirus/2019-ncov/science/science-briefs/scientific-brief-omicron-variant.html, accessed on April 24, 2022.
61 Nyberg T, et al., Comparative analysis of the risks of hospitalisation and death associated with SARS-CoV-2 Omicron (B.1.1.529) and Delta (B.1.617.2) variants in England: A cohort study, *Lancet*, available at www.thelancet.com/journals/lancet/article/PIIS0140-6736(22)00462-7/fulltext, accessed on March 20, 2022.
62 Ibid.

63 Coronavirus Disease 2019 (COVID-19), *BMJ Best Practice*, available at https://best-practice.bmj.com/topics/en-us/3000168/aetiology, accessed on April 20, 2022.

64 Chen J, Omicron variant (B.1.1.529): Infectivity, vaccine breakthrough, and antibody resistance, *J Chem Inf Model*, available at www.ncbi.nlm.nih.gov/pmc/articles/PMC8751645/, accessed on April 27, 2022.

65 Sheikh A, et al., Severity of Omicron variant of concern and effectiveness of vaccine boosters against symptomatic disease in Scotland (EAVEII): A national cohort study with nested test-negative design, *Lancet*, available at www.thelancet.com/journals/laninf/article/PIIS1473-3099(22)00141-4/fulltext, accessed on April 25, 2022.

66 Omicron variant: What you need to know, CDC, available at www.cdc.gov/coronavirus/2019-ncov/variants/omicron-variant.html, accessed on April 24, 2022.

67 Coronavirus disease 2019 (COVID-19), *BMJ Best Practice*, available at https://bestpractice.bmj.com/topics/en-us/3000168/aetiology, accessed on April 20, 2022.

68 Ibid.

69 Nyberg T, et al., Comparative analysis of the risks of hospitalisation and death associated with SARS-CoV-2 Omicron (B.1.1.529) and Delta (B.1.617.2) variants in England: A cohort study, *Lancet*, available at www.thelancet.com/journals/lancet/article/PIIS0140-6736(22)00462-7/fulltext, accessed on March 20, 2022.

70 Ibid.

71 Ibid.

72 Andrews N, Covid-19 vaccine effectiveness against the Omicron (B.1.1.529) variant, *NEJM*, available at www.nejm.org/doi/full/10.1056/NEJMoa2119451#:~:text=After%20a%20BNT162b2%20primary%20course,at%205%20to%209%20weeks, accessed on April 24, 2022.

73 Coronavirus Disease 2019 (COVID-19), *BMJ Best Practice*, available at https://bestpractice.bmj.com/topics/en-us/3000168/aetiology, accessed on April 20, 2022.

74 World Health Organization, Enhancing readiness for Omicron (B.1.1.529): technical brief and priority actions for member states, *WHO*, available at chrome-extension://efaidnbmnnnibpcajpcglclefindmkaj/www.who.int/docs/default-source/coronaviruse/2021-12-23-global-technical-brief-and-priority-action-on-omicron.pdf?sfvrsn=d0e9fb6c_8, accessed on May 4, 2022.

75 Andrews N, Covid-19 vaccine effectiveness against the Omicron (B.1.1.529) variant, *NEJM*, available at www.nejm.org/doi/full/10.1056/NEJMoa2119451#:~:text=After%20a%20BNT162b2%20primary%20course,at%205%20to%209%20weeks, accessed on April 24, 2022.

76 Statement on Omicron sublineage BA.2, WHO, available at www.who.int/news/item/22-02-2022-statement-on-omicron-sublineage-ba.2#:~:text=2%20sublineage%20should%20continue%20to,Omicron%20by%20public%20health%20authorities, accessed on April 24, 2022.

77 Ibid.

78 Elisabeth Mahase, Covid-19: What do we know about Omicron sublineages? *BMJ*, 2022; 376, https://doi.org/10.1136/bmj.o358, available at www.bmj.com/content/376/bmj.o358.long, accessed on March 1, 2022.

79 COVID-19 weekly epidemiological update, WHO, available at https://www.who.int/docs/default-source/coronaviruse/situation-reports/20220201_weekly_epi_update_77.pdf?sfvrsn=f0b4115f_26&download=true, accessed on April 24, 2022.

80 Statement on Omicron sublineage BA.2, WHO, available at www.who.int/news/item/22-02-2022-statement-on-omicron-sublineage-ba.2#:~:text=2%20sublineage%20should%20continue%20to,Omicron%20by%20public%20health%20authorities, accessed on April 24, 2022.

81 UK Health Security Agency, Investigation of SARS-CoV-2 variants: Technical briefing documents on novel SARS-CoV-2 variants, 2022.

82 Statement on Omicron sublineage BA.2, WHO, available at www.who.int/news/item/22-02-2022-statement-on-omicron-sublineage-ba.2#:~:text=2%20sublineage%20should%20continue%20to,Omicron%20by%20public%20health%20authorities, accessed on April 24, 2022.

83 UK Health Security Agency, Investigation of SARS-CoV-2 variants: Technical briefing documents on novel SARS-CoV-2 variants, 2022.

84 Mahase E, Covid-19: What do we know about Omicron sublineages? *BMJ*, 2022; 376, https://doi.org/10.1136/bmj.o358, available at www.bmj.com/content/376/bmj.o358. long, accessed on March 1, 2022.

85 Ibid.

86 What might rising Covid cases in South Africa mean for the UK? *The Guardian*, available at www.theguardian.com/world/2022/may/02/covid-cases-south-africa-omicron-ba4-ba5-uk, accessed on May 8, 2022.

87 Three new Covid Omicron subvariants detected in Australia, *The Guardian*, available at www.theguardian.com/world/2022/may/04/three-new-covid-omicron-subvariant-australia-ba-2-12-1-ba4-ba5-detected, accessed on May 8, 2022.

88 Omicron lineages BA.4 and BA.5 FAQ, NICD South Africa, available at www.nicd. ac.za/omicron-lineages-ba-4-and-ba-5-faq/, accessed on May 8, 2022.

89 Ibid.

90 Ibid.

91 Latest confirmed cases of Covid-19 in South Africa (7 May 2022), NICD South Africa, available at www.nicd.ac.za/latest-confirmed-cases-of-covid-19-in-south-africa-7-may-2022/, accessed on May 8, 2022.

92 Covid-19 weekly epidemiological update, WHO, available at https://www.who.int/docs/default-source/coronaviruse/situation-reports/20220322_weekly_epi_update_84.pdf?sfvrsn=9ec904fc_8&download=true, accessed on April 24, 2022.

93 What you need to know about variants, CDC, available at www.cdc.gov/coronavirus/2019-ncov/variants/about-variants.html, accessed on May 3, 2022.

11

2020 AND BEYOND

The global geographies of COVID-19

The years 2020 and 2021 will be remembered by humanity as a period of panic, death, hopelessness and chaotic and disorganized efforts of public authorities all over the world against the deadly coronavirus. Lack of genuine evidence for formulation of healthcare policies and lack of wholehearted international cooperation were the hallmarks of this period. The pandemic has exposed the fragile global healthcare systems and has highlighted the need to strengthen manifold the healthcare systems of many countries. The pandemic has also exposed the nationalistic movements and selfish attitude of various countries to secure drugs, vaccines and PPE kits.[1]

The COVID-19 pandemic has been the biggest global healthcare challenge in recent times. No single government or healthcare authority can manage this pandemic individually. A robust global healthcare architecture is the need of the hour. The COVID-19 pandemic has given everyone a painful reminder that nobody is safe until everyone is safe.[2]

Latest update on global spread of COVID-19

As on May 27, 2022, 525,467,084 confirmed cases of COVID-19, including 6,285,171 deaths, have been reported to the WHO.[3] However, the global spread of cases has not been uniform; rather the overall pattern so far has been a series of COVID-19 waves: surges in new cases followed by declines. The surges and the declines have been at different times in different countries or geographical area. Broadly globally, the first surge of cases occurred in December 2020 and January 2021 followed by a decline. The second surge occurred in the months of April–May 2021 which was followed by the third surge in the months of July–August 2021. The fourth massive global surge occurred in the month of January 2022 which is declining at the time of writing and hopefully will further decline in coming months.

DOI: 10.4324/9781003345091-11

The top five countries reporting the maximum number of confirmed cases to the WHO are: the USA (82,853,070 cases), India (43,147,530 cases), Brazil (30,846,602 cases), France (28,561,540 cases) and Germany (26,199,643 cases). These countries are closely followed by the UK, Russian Federation, Republic of Korea, Italy, Turkey and Spain.[4] Of all the geographical areas, Europe has reported the maximum number of cases followed by the Americas and South East Asia.

Spread in European region

From the first COVID-19 cases in Europe reported on January 24, 2020, the pandemic reached 1 million cases within three months, 10 million cases within eight months and 100 million cases in Europe alone within two years.[5] The pandemic has profound effect on health and has disproportionately affected the older age group across the European region. In early 2020, people over the age of 50 years accounted for about 70% of cases and close to 100% of deaths. Later the demographics changed and by the end of 2021, nearly 70% of the cases were between the ages of 5 and 49 years, while those over 50 continued to account for the majority of reported deaths.[6]

The trend of infection in European region has been deeply impacted by the emergence of several variants of concern of SARS-CoV-2 that lead to surges of infection. By April 2021, the Alpha variant of concern was dominant in the European region. It was replaced by the Delta variant by August 2021, which led to more deaths throughout the European region. Later it was replaced by the highly transmissible Omicron variant. From September 2021, with the onset of the winter season the cases started increasing in the European region and they are still rising in several countries, mainly caused by the Omicron variant. France, Italy, Greece and Austria have recently seen their daily infections increase.[7] However, daily cases are now lower in many European countries than their previous peaks, though some are seeing a rise driven by an easily spread subvariant of Omicron called BA.2.

The surges of cases were not uniform throughout Europe. In some countries when the cases were increasing, in others they were declining and this trend has persisted throughout the pandemic. In March 2020, the cases overall peaked in Europe and after that there was an overall decline. The cases started rising again in September 2020 and remained high till the first week of April 2021 and then started declining. However, the decline was not uniform in all the countries of Europe. From the third week of June 2021 the cases started climbing again and reached a peak in the middle of January 2022, which was fuelled by the highly transmissible Omicron variant.[8] Cases started rising in the middle of 2021, and by late 2021, Europe had become the epicentre of the pandemic.[9]

As reported by the BBC, on September 18, 2020, Hans Kluge, the WHO regional director, said that in the past two weeks the number of new cases had doubled in more than half of European Member States. He further said that 300,000 new infections were reported across Europe in the previous week alone and weekly cases had exceeded those reported during the first peak in March.[10] On the same day the BBC further reported that Spain had the highest number of coronavirus cases in Europe, and Madrid was once again the worst-hit region and France recorded its highest number of new confirmed daily cases since the pandemic began, at 13,215 – a jump of nearly 3,000 more cases in 24 hours. Several cities, including Marseille and Nice, were bringing in tighter restrictions. Cases were surging in the UK as well, and large parts of the north of England were subjected to extensive lockdown measures.[11] On October 27, 2020, the BBC reported that according to WHO spokeswoman Dr Margaret Harris, France, Spain, the UK, the Netherlands and Russia accounted for the majority of cases which had increased by a third. She further said that, 'Across the European region we're seeing an intense and indeed alarming increase in cases and deaths'. The WHO further informed that Europe's daily Covid deaths rose by nearly 40% compared with the previous week.[12] In November 2020, the situation further worsened in Europe. On November 10, 2020, Reuters reported that Europe represented just over half of all new infections reported globally. France, the worst affected country in the EU, had registered more than 48,700 infections per day over the past week and the Paris region health authority said 92% of the region's ICU capacity was occupied.[13] Britain reported on an average 20,000 cases daily and Italy reported daily average new cases were at a peak at more than 32,500.[14] During this period France, Spain, Italy and Russia had reported hundreds of deaths a day and together, the five countries accounted for almost three quarters of the total fatalities.[15] In the month of December 2020, with few exceptions, such as southern Ireland, Corsica and parts of Greece and Norway, all regions of Europe were on their highest alert levels.[16] Belgium, a country of 11 million people, suffered one of Europe's highest death rates per capita from the pandemic and had imposed tighter control than neighbours such as Germany and the Netherlands, both of which were in full lockdowns, to prevent the spread of the virus in the country.[17] The situation deteriorated in Britain further and on December 20, 2020, Reuters reported that Britain's European neighbours began closing their doors to travellers from the United Kingdom.[18] In January 2021, the situation improved a bit in Europe. The WHO weekly update dated January 24, 2021, reported 1,328,460 new cases in the previous seven days in Europe, a decline of 20% over the previous week.[19] In this week the new cases declined by 24% in the United Kingdom of Great Britain and Northern Ireland and by 9% in the Russian Federation. However, France reported a 10% weekly rise in new cases.[20] The WHO weekly update dated March 14, 2021, reported 1,225,972 new cases in the previous seven days in Europe, an increase of 6% over the previous week.[21] Italy reported a 12% increase in new cases as compared to the previous week and France reported a rise in 5% of new cases. In the WHO report dated May 9, 2021, the situation stabilized to some extent in Europe where the weekly cases showed a 23% decline.[22] Overall the cases declined to a great extent by the first week of June 2021, in the WHO European region.[23]

This wave had hardly declined in the first week of June 2021 when cases started rising again from the last week of June 2021.[24] The new weekly cases showed a marginal decline in the middle of September 2021, only to rise steadily from

October 2021. There was an exponential rise in cases in Europe which peaked in late January 2022.[25] The WHO weekly report dated October 19, 2021, for the European region, reported a 7% increase in the number of new weekly cases as compared to the previous week. The United Kingdom reported a 14% increase in new weekly cases and the Russian Federation reported a rise of 15%.[26] The WHO in November 2021, warned that Europe was again the epicentre of the pandemic.[27] Insufficient vaccine uptake and relaxation of public health measures were to blame for the rise in cases. Germany recorded more than 37,000 daily new cases on November 5, 2021. However, there was a dramatic rise in fatalities in Russia and Ukraine where vaccine coverage had been very low. Cases increased in Romania, Hungary, Croatia, Slovakia, England, Italy and Portugal. However, the number of cases did not rise significantly in Spain.[28] By late November, the WHO regional director Dr Hans Kluge warned that 500,000 more deaths could be recorded by March 2022. He said that factors like the winter season, insufficient vaccine coverage and the regional dominance of the more transmissible Delta variant were behind the spread. Austria became the first European country to announce that COVID-19 vaccination would become a legal requirement to check the alarming spread of COVID-19. However, protest started in Vienna against this new measure. Similarly, violent rioting erupted in Rotterdam, Netherlands over new COVID-19 measures. People opposed new restrictions imposed to prevent the transmission. Germany and UK did not impose any national lockdown.[29] In January 21, 2022, the BBC reported that the WHO warned that half of Europe would catch the Omicron Covid variant within the next six to eight weeks. Dr Hans Kluge said a 'west-to-east tidal wave' of Omicron was sweeping across the region, on top of a surge in the Delta variant.[30] The cases however slowly stabilized across Europe, and the WHO weekly report dated February 22, 2022, revealed a 26% decline in new weekly cases as compared to the previous week.[31] From the middle of March 2022, the number of weekly new cases has been continuously declining, and the trend was continuing in the middle of April in the WHO European region.[32]

Spread in the WHO Region of the Americas and Caribbean

The WHO geographic region Americas reported 156,560,365 confirmed cases of COVID-19 on May 27, 2022, the second highest cumulative case count after Europe.[33] There were several surges of cases in the Americas. The first surge peaked around June–July 2020. The second surge peaked around early January 2021. The third wave peaked around mid-April 2021. The fourth wave peaked around mid-August 2021 and the last massive surge of cases peaked around early January 2022.[34] The USA and Brazil suffered the worst and reported the maximum number of cases from this region.

The cumulative case count in the USA is more than 80 million, the highest figure in the world. However, recently the cases have started declining. COVID-19 cases surged in the US in the last months of Trump's presidency, with North America's case rate peaking at 470 per million people on January 11, 2021.[35] The number of new weekly cases in the United States had started rising since the week beginning September 7, 2020, when there were 240,000 confirmed cases to over 400,000 confirmed cases over the previous week (the WHO report dated October 25, 2020).[36] By the end of November 2020, the USA reported the highest number of weekly cases (1.15 million) to the WHO.[37] By the end of January 2021, the highest number of weekly new cases was still reported by the United States of America (1,072,287 cases); however, there was a decline by 15%.[38]

Another surge of cases occurred in the month of August 2021. The WHO weekly report dated August 24, 2021, reported the highest numbers of new cases from the United States of America (1,020,072 new cases; 308.2 new cases per 100,000; 15% increase).[39] After this surge the number of cases declined. However, the cases started rising again from the first week of December 2021 and they peaked in mid-January 2022 followed by a gradual decline. According to the WHO weekly report dated December 28, 2021, the highest numbers of new cases continued to be reported from the United States of America (1,185,653 new cases; 358.2 new cases per 100,000; a 34% increase), in the region of the Americas.[40] According to the CDC, after detection of the first confirmed case of the Omicron variant in the USA on December 1, 2021, it spread rapidly and by December 20, 2021, Omicron had been detected in every US state and territory and continued to be the dominant variant in the United States.[41] By the last week of February 2022, the United States of America reported 457,058 new cases; 138.1 new cases per 100,000; a decline of 36.[42] In the week ending March 27, 2022, the number of weekly cases had shown a rise of 67.1% in Ohio, 59.1% in Texas and 44.7% in Delaware. New York State had also shown a surge of 29.4% of cases. Overall, there was a total 12.2% decline of cases in the week ending March 27, 2022.[43] Dr Anthony Fauci, the country's chief adviser on infectious disease, has expressed concern that hospitalizations could still put pressure on health services in areas of the country where people had not been fully vaccinated or had a booster.[44]

In Latin America, cases are on decline nearly everywhere after a recent rise in many countries. Brazil, the worst-hit country in the region, recorded more than 29 million cases. The first surge of cases in Brazil happened in the middle of 2020 which peaked in late July 2020. Cases later declined and started gradually increasing from November 2020 to reach a peak around March 2021 and declined by November 2021. The Omicron-fuelled cases surged in the first week of January 2022 and reached a peak by the middle of January 2022. The cases have declined at the time of writing to a great extent.[45]

A heterogeneous evolution in the incidence of COVID-19 has been observed in the Latin American countries. This heterogeneity is associated with both the public health measures adopted, as well as with the population size, poverty levels and pre-existing health systems.[46] By July 2021, the well-known variant of concern, Gamma (P.1), initially found in Brazil, had spread to neighbouring countries and became dominant in Uruguay. Lambda (C.37), first detected in Peru and considered a variant of interest, was responsible for up to 80% of infections in Peru. More worryingly, the highly transmissible Delta (B.1.617.2) variant had also been detected in at least 16 countries in Latin America and Caribbean countries and community transmission had already started by July 2021.[47]

The number of cases declined in this region to some extent in the next few months only to rise again in December 2021, in an unprecedented manner fuelled by the highly infectious Omicron variant. Health services in countries like Brazil, Mexico and Paraguay were stretched to their limit as the second wave hit these countries where the first never really ended.[48] South America has witnessed some of the pandemic's bleakest moments – with bodies dumped in mass graves and patients starved of oxygen in overwhelmed hospitals.[49] At present all the countries in the region are below the peak of their infection curves. Brazil is down 12%, Argentina 2%, Paraguay 1%, Mexico 5% and Chile is 16% down.[50]

Spread in South East Asia

South East Asia suffered three huge surges of COVID-19 cases. The first surge peaked around the middle of September 2020 and it was followed by a huge surge of cases that peaked around May 2021. It was fuelled by the dangerous Delta variant that was first detected in India. This was followed by another surge of cases in January 2022, which has started to decline in the month of March and April 2022. This was fuelled by the highly transmissible Omicron variant.

Overall in Asia the number of cases surged in some countries whereas they declined in some other countries during the last two years. China, which was able to control the initial outbreak in 2019 and early 2020, followed a zero-Covid policy and was mostly successful in controlling the spread of the disease. This has been possible largely due to widespread testing and strict lockdown. However, recently the cases have started going up due to infections caused by the Omicron variant.[51] China reported almost 5,000 COVID-19 cases on March 25, 2022.[52] By far, most of the cases in the ongoing outbreak have been in Jilin province, which borders North Korea in China's industrial northeast. COVID-19 has forced lockdowns in several large cities of China including the financial hub Shanghai this year, making the current outbreak the country's worst since the pandemic started in Wuhan in 2019 and threatening to derail the world's second-largest economy.[53] There have been protests against the harsh measures implemented by the authorities to control the recent surge of cases. Those diagnosed with Covid have been sent to live in warehouses and exhibition halls converted into mass quarantine centres, even if they are asymptomatic, and some have complained about the basic living conditions.[54] The zero-Covid model was strictly enforced both in mainland China and Hong Kong over the last two years and it seemed to be largely paying off. However, the model was put under question when authorities were forced to impose increasingly large lockdowns triggered by the more infectious Delta variant in 2021. Now rising infections triggered by the Omicron variant has placed further question marks over the much vaunted zero-Covid policy.[55]

India was devastated by the COVID-19 pandemic which swept through the country mainly in three successive waves. The first surge peaked around

September 2020. It was followed by the devastating second wave that peaked around early May 2021. The third wave peaked in January 2022. In March 2020, the government of India announced a strict countrywide lockdown to prevent the spread of COVID-19. However, cases started appearing from almost every corner of the vast country. The first wave started from June 2020 and peaked in September 2020 and declined slowly. According to the WHO situation report dated May 31, 2020, there were 89,995 active cases and 86,983 were cured/discharged.[56] ICMR advised States to conduct sero-surveys to measure coronavirus exposure in a selected population using the IgG ELISA Test. A WHO representative had discussions with the Health Secretary and CEO of NITI Ayog and advocated for enhancing the Covid response in densely populated urban areas of India. By June 28, 2020, the cases surged to 203,051 active cases; 309,712 were cured/discharged. PM CARES Fund Trust had allocated Rs. 2000 crore[57] for supply of 50,000 'Made-in-India' ventilators to government-run Covid hospitals in all States/UTs for management of serious cases of COVID-19.[58] Efforts of containment failed to halt the spread of the virus and by July 26, 2020, there were 467,882 active cases[59] which further increased to 710,771 active cases by August 24, 2020.[60] By this time Covid had spread to almost every corner of the country and it peaked in the month of September when the active case count rose to 5,992,532 on September 27, 2020.[61] After this slowly the cases started declining and by December 2020, the number of new cases declined significantly.

The people of India had hardly breathed a sigh of relief after the onslaught of the first Covid wave when the second wave of COVID-19 started in the month of April 2021. On April 9, India surpassed Brazil as having the second most cases of COVID-19 worldwide.[62] By the end of April, India was reporting the highest number of daily cases in the world, almost 50% of new cases reported in the world. Among Indian states Maharashtra was leading the tally with the highest number of cases (more than 4 million) followed by Kerala, Karnataka, Uttar Pradesh and Tamil Nadu (all more than 1 million).[63] The cases started rising alarmingly and the wave peaked in late April and early May 2021. It was caused by the deadly, highly transmissible Delta variant of COVID-19. The virus escaped all the pandemic control measures like isolation, quarantine and lockdown and soon replaced the older variant. On May 7, 2021, India reported 414,188 new cases, the highest ever single day spike.[64] The variant ripped through India in a catastrophic surge in a very short time, killing thousands in matter of weeks. However, by end of May 2021, the number of new cases had started declining.[65] The decline of cases was rapid and by September 2021, the cases reduced to a great extent. The Delta variant caused havoc in India. Entire families were infected, hospitals were overwhelmed.[66]

After the decline of the massive second wave, the people of India were relieved to a great extent and felt that COVID-19 was more or less over. However, the Omicron variant of SARS-CoV-2, caused the third wave in India. It started abruptly at the end of December 2021 and the first week of January 2022 and reached its peak in the third week of January. India reached the peak of the Omicron-driven third wave on January 21, 2022, when it recorded 347,000 new cases.[67] It declined afterwards rapidly and by the middle of February 2022, the surge of cases was almost over. However, the decline was not uniform all over India. In Kerala and Mizoram the number of cases did not decline rapidly as compared to rest of the country. Soumya Swaminathan, Chief Scientist of the WHO warned people not to be complacent about the virus, which still had the potential to overwhelm India's healthcare system, even if it was milder than the Delta variant, adding that it was 'not the common cold'. However, people had thrown caution to the wind and as a result the virus ran amok amongst the careless people. Fortunately, hospitalization was less but the healthcare workers had to bear the brunt. More healthcare workers fell ill with the virus during this wave than before. In Delhi alone, more than 700 healthcare workers in six big government hospitals tested positive.[68] At the time of writing the number of active cases in India is 16,308.[69]

COVID-19 in WHO Africa region

The African region faced four waves of COVID-19. The first wave peaked in July 2020. By the end of May 2020, transmission was present in many countries. However, the panic mode had not set in. The majority of the imported cases arrived from Europe and the USA.[70] From early November 2020, cases started surging again and the second wave peaked around the first week of January 2021.[71] In early June 2021 the third Covid wave started in Africa and it peaked around the middle of July 2021.[72] According to the WHO, in the first week of June the pandemic was trending upwards in 14 countries and eight countries had witnessed an abrupt rise of over 30% in cases.[73] The brutal wave was triggered by the Delta variant. According to the WHO, more than 251,000 new cases were reported in Africa in the week ending July 4, 2021, a 20% increase from the previous week and it was labelled the 'worst pandemic week ever'.[74] The third wave ripped through the countries mainly in southern and eastern Africa. In Uganda, hospitals were stretched thin. Rwanda restricted movement in its capital and Kenya instituted partial lockdowns and extended curfew hours in over a dozen counties where the Delta variant caused huge surges.[75]

The fourth Covid wave started in the second half of November 2021 and it abruptly peaked in the middle of December 2021. This wave was driven by the Omicron and Delta variants. There was an 83% surge in new COVID-19 cases during the second week of December 2021. Africa recorded more than 196,000 new cases for the week ending on December 12, 2021, up from around 107,000 in the previous week, bringing the total cumulative number of recorded cases during the pandemic to 8.9 million. The number of new COVID-19 cases was doubling every five days, the shortest reported in the year 2021.[76] During this wave the Omicron variant was first detected in South Africa and the country had a huge surge of cases. The epicentre of this outbreak was Gauteng, the province that contains Pretoria and Johannesburg. Seventy-four percent of the samples sequenced in last three weeks of November 2021 revealed the presence of the Omicron variant, which showed that this variant was out-competing the Delta variant which had already replaced the Beta variant.[77] By the end of December the cases declined rapidly. The rapid rise in cases was accompanied by a rapid fall. As compared to the previous wave, and as feared, hospitalization, oxygen requirement and death remained low.[78]

At the time of writing 47 countries in the African region had been affected by COVID-19. The total number of cases is 8,525,314.[79] Cases are declining in this region at present. The number of new COVID-19 cases in the WHO African region decreased by 44.0% during the week of March 28 to April 3, 2022, as compared to the previous week. According to the WHO Weekly Bulletin, data as reported by April 3, 2022, in the past seven days, 21 countries (46.0%) reported a decrease of 20% or more in the number of new cases, while Côte d'Ivoire, Eritrea, Eswatini, Mali, Mauritania, Niger, Senegal and Seychelles saw a 20% or more increase in weekly cases compared to the previous week. Most of the new cases were reported from the top five countries (11,421, 89.4%), with South Africa recording the highest number (9,702 new cases, 7.4% increase, 16 new cases per 100,000 population), followed by Zimbabwe (705 new cases, 46% decrease, 9.0 new cases per 100,000), Zambia (481 new cases, 27.0% decrease, 2.5 new cases per 100,000), Seychelles (335 new cases, 43.0% increase, 337.0 new cases per 100,000) and Ethiopia (198 new cases, 19.0% increase, 0.2 new cases per 100,000).[80] Only one country, Mauritius, met the criteria for resurgence (a 20% increase in new COVID-19 cases for at least two consecutive weeks) where cases in the previous week reached 30% or more of the country's highest weekly number of cases. South Africa has recorded the highest number of COVID-19 cases in the region, with 3,722,954 cases (45.3% of all reported cases).[81]

The reported case counts, however, underestimate the overall burden of COVID-19, as only a fraction of acute infections are diagnosed and reported. Besides these, many patients were undiagnosed as they were asymptomatic. Seroprevalence surveys in the United States[82] and Europe[83] have suggested that the actual case count exceeds the incidence of reported cases by several fold.

Global mortality due to COVID-19

As on May 27, 2022, the WHO has reported that 6,285,171 people have died due to COVID-19 in the ongoing pandemic all over the world. The Global death rate increased with the increase in the number of cases and fell with the decline of cases.[84] The death toll has been high in the badly affected countries. In general, the number of cases and number of deaths have gone hand in hand. There was a proportionate increase in the number of cases and deaths globally till the last global wave triggered by the Omicron variant. The number of cases was very high and increased rapidly throughout the globe; however, there was not a proportionate rise in the number of deaths during the Omicron-fuelled global surge.[85]

According to the WHO report published on May 27, 2022, the USA has reported 996,108 deaths; India has reported 524,539 deaths; Brazil has reported 666,037 deaths; France has reported 144,809 deaths and Germany has reported 138,781 deaths.[86] Further analysis of COVID-19 deaths has revealed that 312.44

persons out of 100,000 population have died in Brazil. Similarly in Poland (306.76), in the USA (302.59), in Chile (301.30) and in Italy 271.54 persons have died out of 100,000 population. The death toll in Peru has been 645.72 per 100,000 population, probably the highest in the world. The case fatality ratio (number of deaths divided by number of cases) has been highest in Mexico (5.7%) followed by Indonesia (2.6%), Brazil (2.2%), Russia (2.1%) and Poland (1.9%). Both in the USA and India, which together contribute to a huge number of Covid deaths, the case fatality rates have been 1.2%.[87]

The mortality figures, however, do not reflect the true picture of panic, pain, emotional turmoil, depression and devastation of life suffered by human beings during this pandemic. Overnight, life turned upside down for many families and individuals throughout the world. The first casualty of the COVID-19 pandemic for patients nearing death and even after death is human dignity. The patients were left alone in the hospital; both family members and healthcare workers were afraid of catching infection. As a result, medical care had changed in a way not seen for decades in modern times. The human contact of the dying patients with a healthcare team was only through several layers of gloves, goggles, a face shield and PPE kit. Video calls became the only means of communication with family members for a fortunate few. Patients kept in isolation went through emotional turmoil and suffered from loneliness, isolation, depression and anxiety before their death.[88] Family members suffered in equal amount, even denied the opportunity to say final goodbyes. Only a few received a dignified funeral service. For the majority it was a hurriedly conducted service, mostly in absence of any family member or only a few close family members. The family members had very little opportunity to realize the reality of death and to honour the life of the deceased. The funeral rites carried less priority than the need to contain the spread of COVID-19. The family members were torn between the urge of providing care to the gravely sick family member and the urge for self-protection.

An emergency national law in March 2020 banned funeral services in Italy to prevent the spread of the virus.[89] The haunting memories of images of military trucks transporting piled-up coffins out of Bergamo last year are still raw.[90] In May 2021, New York City was still using refrigerated trucks to store bodies of COVID-19 victims, more than a year after they were first set up as temporary morgues as deaths surged at the height of the pandemic.[91] In July 2020, the BBC reported that in Guayaquil, Ecuador the city's health services began to collapse under the weight of COVID-19 deaths, and hospitals, morgues and cemeteries became overwhelmed. The bodies of many victims of COVID-19 were left uncollected, with people burying loved ones in back gardens or leaving coffins on the streets.[92] In April 2021, in Brazil, bodies had to be exhumed to make way for new dead bodies as the number of deaths exponentially increased, overwhelming healthcare and funeral services.[93] Thousands of old people died in care homes in France, UK, Spain, Italy, Germany and other countries of Europe due to COVID-19.[94] In June 2021, the BBC reported that in many African countries there was a crisis of severe oxygen shortage leading to preventable COVID-19 deaths.[95] The worst possible scenario of COVID-19 deaths was seen in India during the deadly Delta-variant-fuelled Covid surge in April–May 2021. The BBC in April 2021 reported that many patients were dying due to lack of oxygen in the capital New Delhi and several hospitals were forced to stop admissions due to lack of oxygen.[96] The condition did not improve even in May 2021. Patients, including a senior doctor, died due to lack of oxygen in a reputed private hospital in Delhi. Desperate, panic-stricken people were seen rushing patients from one hospital to another with the faint hope of getting a bed equipped with oxygen. People were seen queuing up for hours in the scorching sun to get the cylinders refilled with oxygen. Many charitable institutions opened up facilities to provide oxygen to the needy. However, thousands of people could not make it and many died at home and many died waiting outside the emergency department of hospitals. The crematoriums were running day and night in Delhi and many other cities of India.[97] Hundreds of dead bodies were seen floating in the river Ganga in North India, presumed to be left in the river without proper funerals. On May 10, 2021, 71 corpses washed up on the river bank in the state of Bihar in North India. The bodies on the river banks, taken together with funeral pyres burning around the clock and cremation grounds running out of space, tell the story of a death toll unseen and unheard of in the recent history of modern India.[98]

During this ongoing pandemic the accurate calculation of mortality data is very difficult, and what we get from all reliable sources is at best an excellent estimate of the true picture. The reasons are multiple. Inadequate testing facilities, hesitancy in testing, lack of data collection facilities due to poor public health infrastructure, poor public health policies are some of the important limiting factors.[99] Even the WHO has acknowledged that the percentage of registered deaths ranged from 98% in the European region to only 10% in the African region.[100] The method of measuring deaths during COVID-19 was not uniform. Three main methods used to measure deaths were deaths within 28 days of a positive test result, death certificates mentioning COVID-19 as the cause of death and deaths over and above the usual number at the time of year.[101] A recent report published in the journal *Lancet* has revealed that there is an estimated 18.2 million deaths worldwide (as measured by excess mortality) from January 1, 2020, to December 31, 2021, due to COVID-19. The number of excess deaths due to COVID-19 was largest in the regions of South Asia, north Africa and the Middle East, and eastern Europe. The top three countries with the maximum number of cumulative excess deaths due to COVID-19 were estimated in India (4.07 million), the USA (1.13 million) and Russia (1.07 million). Out of these countries the excess mortality rate was highest in Russia (374.6 deaths per 100,000 people).[102]

On May 5, 2022, the WHO published the data on excess mortality. It reported that the full death toll associated directly or indirectly with the COVID-19 pandemic (described as 'excess mortality') between January 1, 2020, and December 31, 2021, was approximately 14.9 million (of a range of 13.3 million to 16.6 million). Excess mortality includes deaths associated with COVID-19 directly (due to the disease) or indirectly (due to the pandemic's impact on health systems and society). Eighty-four percent of the excess deaths are concentrated in South-East Asia, Europe and the Americas. Some 68% of excess deaths are concentrated in just ten countries globally.[103] Dr Tedros Adhanom Ghebreyesus, WHO Director-General, said that,

> These sobering data not only point to the impact of the pandemic but also to the need for all countries to invest in more resilient health systems that can sustain essential health services during crises, including stronger health information systems.

The full magnitude of this ongoing pandemic will probably never be known in the future as a majority of the countries need to further strengthen death registration and reporting systems besides overcoming the political barrier to accurate reporting. Accurate tracking and monitoring is the need of the hour to efficiently prevent and manage future pandemics. We must remember that the pandemic is not yet over.

Notes

1 Taghizade S, et al., Covid-19 pandemic as an excellent opportunity for global health diplomacy, available at www.frontiersin.org/articles/10.3389/fpubh.2021.655021/full, accessed on April 2, 2022.
2 COVID-19 shows why united action is needed for more robust international health architecture, available at www.who.int/news-room/commentaries/detail/op-ed–covid-19-shows-why-united-action-is-needed-for-more-robust-international-health-architecture, accessed on April 2, 2022.
3 WHO coronavirus (COVID-19) dashboard, available at https://covid19.who.int/, accessed on May 28, 2022.
4 WHO coronavirus (COVID-19) dashboard, available at https://covid19.who.int/table, accessed on May 28, 2022.
5 One hundred million cases in one hundred week: working towards better COVID-19 outcomes in the WHO European Region (2022), available at www.euro.who.int/en/health-topics/health-emergencies/coronavirus-covid-19/publications-and-technical-guidance/2022/one-hundred-million-cases-in-one-hundred-week-working-towards-better-covid-19-outcomes-in-the-who-european-region-2022, accessed on April 3, 2022.
6 Ibid.
7 Covid map: Coronavirus cases, deaths, vaccinations by country, *BBC*, available at www.bbc.com/news/world-51235105, accessed on April 3, 2022.
8 WHO, available at https://covid19.who.int/, accessed on April 15, 2022.
9 Covid-19's epicentre again: Europe faces fresh reckoning, available at www.aljazeera.com/news/2021/11/12/covid-19s-epicentre-again-europe-faces-fresh-reckoning, accessed on April 15, 2022.
10 Covid-19: New fear grips Europe as cases top 30m worldwide, *BBC*, available at www.bbc.com/news/world-54199825, accessed on April 15, 2022.
11 Covid-19: Lockdown in parts of Madrid amid virus spike, *BBC*, available at www.bbc.com/news/world-europe-54211361, accessed on April 15, 2022.
12 Coronavirus: Europe's daily deaths rise by nearly 40% compared with last week – WHO, *BBC*, available at www.bbc.com/news/world-europe-54704677, accessed on April 15, 2022.

13 Shaina Ahluwalia, et al., Europe COVID death toll set to pass 300,000 as winter looms and infections surge, *Reuters*, available at www.reuters.com/article/health-coronavirus-europe-idUSL1N2HW099, accessed on April 15, 2022.

14 Ibid.

15 Ibid.

16 Belgium tightens COVID-19 controls with tests for all travelers, *Reuters*, available at www.reuters.com/article/us-health-coronavirus-belgium-idUSKBN2941ZU, accessed on April 15, 2022.

17 Ibid.

18 European neighbours shut doors to Britain amid alarm over new coronavirus strain, *Reuters*, available at www.reuters.com/business/healthcare-pharmaceuticals/european-neighbours-shut-doors-britain-amid-alarm-over-new-coronavirus-strain-2020–12–20/, accessed on April 15, 2022.

19 WHO Covid-19 weekly epidemiological update, available at https://www.who.int/docs/default-source/coronaviruse/situation-reports/20210127_weekly_epi_update_24.pdf, accessed on April 15, 2022.

20 Ibid.

21 WHO Covid-19 weekly epidemiological update, available at https://www.who.int/docs/default-source/coronaviruse/situation-reports/20210316_weekly_epi_update_31.pdf?sfvrsn=c94717c2_17&download=true, accessed on April 15, 2022.

22 WHO Covid-19 weekly epidemiological update, available at https://www.who.int/docs/default-source/coronaviruse/situation-reports/20210511_weekly_epi_update_39.pdf?sfvrsn=b66ba70d_11&download=true, accessed on April 15, 2022.

23 WHO, available at https://covid19.who.int/, accessed on April 15, 2022.

24 Ibid.

25 Ibid.

26 WHO Covid-19 weekly epidemiological update, available at https://www.who.int/docs/default-source/coronaviruse/situation-reports/20211019_weekly_epi_update_62.pdf?sfvrsn=f0a4a5fe_27&download=true, accessed on April 15, 2022.

27 Covid: WHO warns Europe once again at epicentre of pandemic, *BBC*, available at www.bbc.com/news/world-europe-59160525, accessed on April 15, 2022.

28 Ibid.

29 Covid: WHO says it is very worried about Europe surge, *BBC*, available at www.bbc.com/news/world-europe-59358074, accessed on April15, 2022.

30 Covid: Half of Europe to be infected with Omicron within weeks – WHO, *BBC*, available at www.bbc.com/news/world-europe-59948920, accessed on April 15, 2022.

31 WHO Covid-19 weekly epidemiological update, available at https://www.who.int/docs/default-source/coronaviruse/situation-reports/20220222_weekly_epi_update_80.pdf?sfvrsn=31931200_3&download=true, accessed on April 15, 2022.

32 WHO, available at https://covid19.who.int/, accessed on April 15, 2022.

33 WHO, available at https://covid19.who.int/, accessed on May 28, 2022.

34 Ibid.

35 Two years of coronavirus: How pandemic unfolded around the world, *The Guardian*, available at www.theguardian.com/world/2021/dec/31/two-years-of-coronavirus-how-pandemic-unfolded-around-the-world, accessed on April 3, 2022.

36 WHO Covid-19 weekly epidemiological update, available at https://www.who.int/docs/default-source/coronaviruse/situation-reports/weekly-epi-update-11.pdf, accessed on April 17, 2022.

37 WHO Covid-19 weekly epidemiological update, available at https://www.who.int/docs/default-source/coronaviruse/situation-reports/20201201_weekly_epi_update_16.pdf?sfvrsn=a731dd9b_13&download=true, accessed on April 17, 2022.

38 WHO Covid-19 weekly epidemiological update, available at https://www.who.int/docs/default-source/coronaviruse/situation-reports/20210202_weekly_epi_update_25.pdf?sfvrsn=b38d435c_4&download=true, accessed on April 17, 2022.

39 WHO Covid-19 weekly epidemiological update, available at https://www.who. int/docs/default-source/coronaviruse/situation-reports/20210824_weekly_epi_ update_54.pdf?sfvrsn=7356167a_7&download=true, accessed on April 17, 2022.

40 WHO Covid-19 weekly epidemiological update, available at https://www.who. int/docs/default-source/coronaviruse/situation-reports/20211228_weekly_epi_ update_72.pdf?sfvrsn=7a3567d4_3&download=true, accessed on April 17, 2022

41 Omicron variant: What you need to know, CDC, available at www.cdc.gov/ coronavirus/2019-ncov/variants/omicron-variant.html, accessed on April 17, 2022.

42 WHO Covid-19 weekly epidemiological update, available at https://www.who. int/docs/default-source/coronaviruse/situation-reports/20220301_weekly_epi_ update_81.pdf?sfvrsn=7632e25_4&download=true, accessed on April 17, 2022.

43 Tracking coronavirus vaccinations and outbreaks in the U.S., *Reuters*, available at https:// graphics.reuters.com/HEALTH-CORONAVIRUS/USA-TRENDS/dgkvlgkrkpb/, accessed on April 3, 2022.

44 Covid map: Coronavirus cases, deaths, vaccinations by country, *BBC*, available at www. bbc.com/news/world-51235105, accessed on April 3, 2022.

45 https://covid19.who.int/region/amro/country/br, accessed on April 18, 2022.

46 Laura Débora Acosta, Response capacity to the COVID-19 pandemic in Latin America and the Caribbean, available at www.paho.org/journal/en/articles/response-capacity-covid-19-pandemic-latin-america-and-caribbean, accessed on April 3, 2022.

47 Covid-19 in Latin America – emergency and opportunity, *The Lancet*, July 10, 2021, available at www.thelancet.com/journals/lancet/article/PIIS0140-6736(21)01551-8/ fulltext, accessed on April 3, 2022.

48 'Just unimaginable': Latin America's Covid crisis lurches from bad to worse, *The Guardian*, available at www.theguardian.com/world/2020/dec/10/latin-america-covid-crisis-heads-from-bad-to-worse, accessed on April 3, 2022.

49 Ibid.

50 Latin America and the Caribbean, *Reuters*, available at https://graphics.reuters.com/ world-coronavirus-tracker-and-maps/regions/latin-america-and-the-caribbean/, accessed on April 3, 2022.

51 Covid map: Coronavirus cases, deaths, vaccinations by country, *BBC*, available at www. bbc.com/news/world-51235105, accessed on April 4, 2022.

52 Frustration with Covid response grows in China as daily cases near 5,000, *The Guardian*, available at www.theguardian.com/world/2022/mar/25/frustration-with-chinas-covid-response-grows-as-daily-cases-near-5000, accessed on April 4, 2022.

53 Covid lockdown extended in Shanghai as outbreaks put economy on the skids, *The Guardian*, available at www.theguardian.com/world/2022/apr/01/covid-lockdown-extended-for-parts-of-shanghai-as-city-struggles-to-control-omicron, accessed on April 4, 2022.

54 China: Panic buying in divided Shanghai under lockdown, *The BBC*, available at www. bbc.com/news/world-asia-china-60912846, accessed on April 4, 2022.

55 Omicron vs zero-Covid: How long can China hold on? *The BBC*, available at www.bbc. com/news/world-asia-china-60762032, accessed on April 4, 2022.

56 India situation report, WHO, available at https://cdn.who.int/media/docs/default-source/wrindia/situation-report/india-situation-report-18.pdf?sfvrsn=7c00a3f_2, accessed on April 5, 2022.

57 One crore is equal to ten million.

58 India situation report, WHO, available at https://cdn.who.int/media/docs/default-source/wrindia/situation-report/india-situation-report-22.pdf?sfvrsn=c49bf98d_2, accessed on April 5, 2022.

59 India situation report, WHO, available at https://cdn.who.int/media/docs/default-source/wrindia/situation-report/india-situation-report-26.pdf?sfvrsn=a292c9c5_2, accessed on April 5, 2022.

60 India situation report, WHO, available at https://cdn.who.int/media/docs/default-source/wrindia/situation-report/india-situation-report-30.pdf?sfvrsn=44654284_2, accessed on April 5, 2022.

61 India situation report, WHO, available at https://cdn.who.int/media/docs/default-source/wrindia/situation-report/india-situation-report-35.pdf?sfvrsn=22c1fe2d_2, accessed on April 5, 2022.

62 COVID-19: India overtakes Brazil with second highest number of cases, available at www.newindianexpress.com/nation/2021/apr/12/covid-19-india-overtakes-brazil-with-second-highest-number-of-cases-2289126.html, accessed on April 5, 2022.

63 India situation report, WHO, available at https://cdn.who.int/media/docs/default-source/wrindia/situation-report/india-situation-report-65.pdf?sfvrsn=712919f8_4, accessed on April 5, 2022.

64 News headlines for May 7, 2021: India's highest single-day COVID tally, available at www.timesnownews.com/india/article/news-headlines-for-may-7-2021-india-s-highest-single-day-covid-tally-mk-stalin-s-oath-as-tn-cm-other-news/754186, accessed on April 7, 2022.

65 India situation report, WHO, available at https://cdn.who.int/media/docs/default-source/wrindia/situation-report/india-situation-report-69.pdf?sfvrsn=9bbf5985_6, accessed on April 5, 2022.

66 India's brutal Delta-driven wave ended when cases suddenly plunged. Weeks later, the stubborn variant persists, *The Washington Post*, available at www.washingtonpost.com/world/2021/08/05/india-delta-variant/, accessed on April 7, 2022.

67 Omicron peaked on January 21 with 3,47,000 daily cases, *The Economic Times*, available at https://economictimes.indiatimes.com/news/india/omicron-peaked-on-january-21-with-347000-daily-cases/articleshow/89335010.cms, accessed on April 8, 2022.

68 What we know so far about India's third wave – and why it's a mistake to dismiss it as 'mild', *Wire*, available at https://thewire.in/health/covid-19-india-third-wave-omicron, accessed on April 8, 2022.

69 www.mygov.in/covid-19, accessed on May 28, 2022.

70 Africa braces for coronavirus, but slowly, *The New York Times*, available at www.nytimes.com/2020/03/17/world/africa/coronavirus-africa-burkina-faso.html, accessed on April 8, 2022.

71 WHO Covid-19 dashboard, available at https://covid19.who.int/, accessed on April 9, 2022.

72 Ibid.

73 Third wave sweeps across Africa as Covid vaccine imports dry up, *The Guardian*, available at www.theguardian.com/world/2021/jun/07/third-wave-sweeps-across-africa-as-covid-vaccine-imports-dry-up, accessed on April 9, 2022.

74 Africa marks its 'worst pandemic week' yet, with cases surging and vaccine scarce, the W.H.O. says, *The New York Times*, available at www.nytimes.com/2021/07/08/world/africa-coronavirus-cases-who.html, accessed on April 9, 2022.

75 Ibid.

76 Africa clocks fastest surge in COVID-19 cases this year, but deaths remain low, WHO, available at www.afro.who.int/news/africa-clocks-fastest-surge-covid-19-cases-year-deaths-remain-low, accessed on April 9, 2022.

77 Owen Dyer, Covid-19: South Africa's surge in cases deepens alarm over Omicron variant, *BMJ*, available at www.bmj.com/content/375/bmj.n3013, accessed on April 9, 2022.

78 Africa clocks fastest surge in COVID-19 cases this year, but deaths remain low, WHO, available at www.afro.who.int/news/africa-clocks-fastest-surge-covid-19-cases-year-deaths-remain-low, accessed on April 9, 2022.

79 www.afro.who.int/health-topics/coronavirus-covid-19, accessed on May 28, 2022.

80 Ibid.

81 Ibid.

82 Havers FP, et al., Seroprevalence of antibodies to SARS-CoV-2 in 10 sites in the United States, March 23–May 12, 2020, *JAMA Intern Med*, 2020; 180(12): 1576–1586.

83 Stringhini S, et al., Seroprevalence of anti-SARS-CoV-2 IgG antibodies in Geneva, Switzerland (SEROCoV-POP): A population-based study, *Lancet*, 2020; 396(10247): 313, available at www.thelancet.com/journals/lancet/article/PIIS0140-6736(20)31304-0/fulltext, accessed on May 4, 2022.

84 Available at https://covid19.who.int/, accessed on May 28, 2022.

85 Available at https://covid19.who.int/, accessed on April 19, 2022.

86 WHO coronavirus (COVID-19) dashboard, available at https://covid19.who.int/, accessed on May 28, 2022.

87 Mortality analysis, available at https://coronavirus.jhu.edu/data/mortality, accessed on May 7, 2022.

88 Chochinov HM, et al., Death, dying, and dignity in the time of the COVID-19 Pandemic, *J Palliat Med*, 2020; 23(10): 1294–1295.

89 Coronavirus: How Covid-19 is denying dignity to the dead in Italy, *BBC*, available at www.bbc.com/news/health-52031539, accessed on April 19, 2022.

90 In Bergamo, memory of coffin-filled trucks still haunts, *France 24*, available at www.france24.com/en/live-news/20210318-in-bergamo-memory-of-coffin-filled-trucks-still-haunts, accessed on April 19, 2022.

91 Bodies of 750 Covid-19 victims in New York City remain in refrigerated trucks, *NBC News*, available at www.nbcnews.com/news/us-news/bodies-750-covid-19-victims-new-york-city-remain-refrigerated-n1266762, accessed on April 19, 2022.

92 Coronavirus: 'When coffins lined the streets of my hometown', *BBC*, available at www.bbc.com/news/av/world-latin-america-53451703, accessed on April 19, 2022.

93 Brazil exhuming bodies to make space in cemeteries for the COVID-19 dead, available at https://nationalpost.com/news/world/a-biological-fukushima-brazil-covid-19-deaths-on-track-to-pass-worst-of-u-s-wave, accessed on April 19, 2022.

94 Coronavirus: The grim crisis in Europe's care homes, *BBC*, available at www.bbc.com/news/world-europe-52094491, accessed on April 19, 2022.

95 African Covid patients 'dying from lack of oxygen', *BBC*, available at www.bbc.com/news/world-africa-57501127, accessed on April 19, 2022.

96 India Covid: Patients dying without oxygen amid Delhi surge, *BBC*, available at www.bbc.com/news/56876695, accessed on April 19, 2022.

97 India Covid: Delhi hospitals plead for oxygen as more patients die, *BBC*, available at www.bbc.com/news/world-asia-india-56940595, accessed on April 19, 2022.

98 Covid-19: India's holiest river is swollen with bodies, *BBC*, available at www.bbc.com/news/world-asia-india-57154564, accessed on April 19, 2022.

99 Understanding data during a pandemic, available at https://datadrivendetroit.org/blog/2020/04/08/understanding-data-during-a-pandemic/, accessed on April 19, 2022.

100 The true death toll of Covid-19, WHO, available at www.who.int/data/stories/the-true-death-toll-of-covid-19-estimating-global-excess-mortality, accessed on April 19, 2022.

101 Covid-19: UK sees over 80,000 excess deaths during pandemic, *BBC*, available at www.bbc.com/news/health-55411323, accessed on April 19, 2022.

102 Wang H, et al., Estimating excess mortality due to Covid-19 pandemic: A systematic analysis of Covid-19 related mortality, 2020–21, *Lancet*, available at www.thelancet.com/journals/lancet/article/PIIS0140-6736(21)02796-3/fulltext, accessed on March 20, 2022.

103 14.9 million excess deaths associated with the COVID-19 pandemic in 2020 and 2021, WHO, available at www.who.int/news/item/05-05-2022-14.9-million-excess-deaths-were-associated-with-the-covid-19-pandemic-in-2020-and-2021, accessed on May 6, 2022.

12

FEATURES, TREATMENTS AND VACCINES

New developments and configurations

In last two years there has been a lot of progress in understanding the clinical manifestations, treatment and vaccine development for COVID-19. It is a new disease and our understanding continues to evolve with time. Science has demystified the disease to a great extent but a lot more is yet to be known to devise an effective preventive and therapeutic strategy to stop the onward relentless progress of the killer virus that has taken a huge toll on humankind. A brief summary of these aspects is covered here.

Clinical manifestations of COVID-19

The clinical spectrum of COVID-19 is very wide. It varies from asymptomatic forms to clinical illness characterized by acute respiratory failure requiring mechanical ventilation, septic shock,[1] and multiple organ failure. The vast majority of symptomatic patients commonly present with fever, cough and breathlessness and less commonly with a sore throat, loss of smell, loss of taste, loss of appetite, nausea, malaise, body ache and diarrhoea.[2] These common features have been observed since the initial days of the pandemic.

It has been observed that the prevalence of fever is higher in adults compared with children. Approximately 54% of children do not exhibit fever as an initial presenting symptom.[3] A cough, a common feature, has been observed to resolve within a week but may persist for more than a few weeks or months.[4] Loss of sense of smell has been a good predictor of infection.[5] Loss of sense of smell and taste may be the only symptoms of COVID-19 in some patients.

Although COVID-19 mainly affects the respiratory system, the other organ systems are also affected.

The prevalence of common gastrointestinal symptoms is as follows: loss of appetite 22.3%; nausea/vomiting 9%; abdominal pain 6.2% and diarrhoea 2.4%.[6] Loss

DOI: 10.4324/9781003345091-12

of appetite and diarrhoea, when combined with loss of smell/taste and fever, were found to be 99% specific for COVID-19.[7] Diarrhoea in children has been associated with a severe clinical course.[8]

The pooled prevalence of common neurologic symptoms like anxiety, depression and insomnia has been reported to be 15.2%, 16% and 23.9%, respectively.[9] Headache, stroke, impairment of consciousness, seizure disorder are some of the important neurologic manifestations.

The commonest eye symptoms reported are dry eye or foreign body sensation (16%), redness (13.3%), tearing (12.8%), itching (12.6%), eye pain (9.6%) and discharge (8.8%). Conjunctivitis was the most common eye disease in patients with eye involvement (88.8%).[10]

Overall prevalence of skin lesions is only 5.7%. However, they may be the only or the first presenting sign. Skin eruptions and rashes are common manifestations.[11] According to one study lesions resembling pseudo-chilblains (40.4%) were the commonest skin manifestations noted in patients with COVID-19.[12] Urticaria has also been observed in COVID-19.

Acute kidney injury is frequently seen in admitted severely ill COVID-19 patients. There is progressive reduction followed by complete loss of urine output. It is associated with increased mortality.[13]

Heart involvement is not uncommon in COVID-19. Myocardial infarction, myocarditis,[14] abnormal rhythms of the heart and shock have been observed in these patients. Besides these, pre-existing cardiovascular disease has been observed to be significantly associated with higher mortality and ICU admission.[15]

Some patients with COVID-19 may have additional infections that are noted when they present for care or that develop during the course of treatment which may complicate treatment and recovery. At the time of presentation, concomitant viral infections, including influenza and other respiratory viruses and community-acquired bacterial pneumonia, have been reported.[16] Hospitalized patients with COVID-19 may develop common hospital-acquired bacterial and fungal infections which make treatment difficult. Opportunistic fungal infections, especially mucormycosis,[17] has been reported from India which develop due to an unholy intersection of a trinity of COVID-19, diabetes and rampant use of corticosteroids.[18]

Classification of COVID-19 patients

The National Institutes of Health (NIH) issued guidelines that classify COVID-19 into five distinct types, based on severity of disease:

Asymptomatic or pre-symptomatic infections

This type includes individuals with a positive SARS-CoV-2 test using a virologic test (NAAT)[19] or antigen test[20] without any clinical symptoms consistent with COVID-19. It is unclear what percentage of individuals who present with asymptomatic infection progress to clinical disease.

Mild illness

This category includes individuals who have any symptoms of COVID-19 such as fever, cough, sore throat, malaise, headache, muscle pain, nausea, vomiting, diarrhoea, loss of smell or taste but without shortness of breath or abnormal chest imaging. Most mildly ill patients can be managed in an ambulatory setting or at home through telemedicine or telephone visits. Older patients and those with underlying co-morbidities are at higher risk of disease progression.

Moderate illness

This includes individuals who have clinical symptoms or radiologic evidence of lower respiratory tract disease and who have oxygen saturation (SpO_2)[21] ≥94% on room air.

Severe illness

Individuals who have (SpO_2) <94% on room air; with respiratory frequency >30 breaths/min or lung infiltrates >50% are categorized as those with severe illness. These patients often deteriorate rapidly and require oxygen therapy.

Critical illness

These individuals suffer from acute respiratory failure, septic shock and/or multiple organ dysfunction. These patients require admission in an intensive care unit and proper care.

Involvement of lungs is common in COVID-19. Pneumonia is one of the main features of COVID-19. Both the lungs are usually involved. The lungs become inflamed and get filled with fluid, leading to breathing difficulties. Air sacs in the lungs get filled with fluid which limits their ability to take in oxygen and in turn causes shortness of breath, cough and other symptoms. A CT scan of the lungs helps in making an early diagnosis.

Patients with critical COVID-19 illness often develop of acute respiratory distress syndrome (ARDS) which tends to occur approximately one week after the onset of symptoms. In this condition the oxygen saturation falls to a critical level as the lungs get filled with fluid. Breathing becomes progressively difficult and mechanical ventilation becomes mandatory. ARDS is one of the commonest causes of death in hospitalized COVID-19 patients.[22]

Patients with certain underlying co-morbidities like age ≥65 years; having cardiovascular disease, chronic lung disease, sickle cell disease, diabetes, cancer, obesity or chronic kidney disease; being pregnant; being a cigarette smoker; being a transplant recipient; and receiving immunosuppressive therapy are at a higher risk of progressing to severe COVID-19.[23]

COVID-19 and pregnancy

The classification for the severity of illness categories listed earlier also applies to pregnant patients as well. Studies have shown an increased risk of developing severe COVID-19 if they are infected, compared with non-pregnant women of a similar age, especially in the third trimester.[24] The Delta variant seems to be associated with more severe disease. The Omicron variant may be associated with less severe disease than the Delta variant, but it is more infectious. There is no reported increase in congenital anomalies incidence because of COVID-19 infection. COVID-19 during pregnancy has also been associated with an increased likelihood of preterm birth.

Transmission of the virus from mother to foetus in the womb or during birth is possible, but very rare.[25] Most babies won't develop COVID-19, and those who develop symptoms tend to recover quickly. Babies can be infected after birth, so all precautions to reduce the risks of passing the virus to the baby need to be taken.[26] WHO recommends that newborn babies can be placed skin-to-skin and breastfed if the mother is confirmed or suspected to have COVID-19. The numerous benefits of skin-to-skin contact and breastfeeding substantially outweigh the potential risks of transmission and illness associated with COVID-19.[27]

Royal College of Obstetrics and Gynaecology has observed that in women with symptomatic COVID-19, there may be an increased risk of foetal compromise in active labour and of caesarean birth. Women with symptomatic suspected or confirmed COVID-19 should be advised to labour and give birth in an obstetric unit with continuous electronic foetal monitoring.[28] Oxygen supplementation is recommended for pregnant patients when SpO_2 falls below 95% on room air at sea level to ensure adequate oxygen delivery to the foetus.[29]

All pregnant women should limit in-person interactions with people who might have been exposed to COVID-19, including people within the household, as much as possible. The American College of Obstetricians and Gynecology strongly recommends COVID-19 vaccination for women who are pregnant, breastfeeding or planning to get pregnant.[30] The CDC recommends a booster dose as well for them.[31]

COVID-19 in children

Younger children, school children and adolescents usually have fewer and milder symptoms of SARS-CoV-2 than adults and are less likely than adults to experience severe COVID-19.[32] The difference in functioning and maturity of the immune system in children and adults is the most probable reason for this. Although there is some evidence that older children have higher rates of asymptomatic disease than infants (< 1 year), the majority of children present with symptomatic disease and do not appear to be silent spreaders of infection.[33]

Although COVID-19 infection in children is usually mild to moderate in nature, a hyperinflammatory syndrome, called paediatric inflammatory multisystem syndrome, temporally associated with SARS-CoV-2 (PIMS-TS) in Europe and multisystem inflammatory syndrome in children (MIS-C) in the USA, can complicate recovery from COVID-19.[34] In this condition different body organs can become inflamed, including the heart, lungs, kidneys, brain, skin, eyes or gastrointestinal organs. MIS-C can be serious, even deadly, but most children who were diagnosed with this condition have become better with medical care.[35]

A WHO scientific brief[36] has published that overall children aged <10 years were less susceptible to infection than older children and adults, although seroprevalence in adolescents appears similar to adults.[37] The scientific brief further recommends that appropriate preventive measures, including physical distancing, cleaning of hands, coughing into a bent elbow or a tissue, adequate ventilation in indoor settings, and masks, should be consistently implemented in schools for all ages, especially since children under the age of 12 years are not yet eligible for vaccination in most contexts.

A study published recently in the medical journal *Lancet* has shown that children are overall not becoming seriously unwell with COVID-19,[38] and data from England show that children are also not requiring intensive care in large numbers.[39] However, overall evidence indicates that children continue to be mostly, but not completely, spared the worst outcome of the pandemic, particularly compared with older adults who have been much harder hit.[40]

Long Covid

The majority who develop COVID-19 fully recover, but about 10–20% of people experience a variety of mid- and long-term effects after they recover from their initial illness. These mid- and long-term effects are collectively known as post COVID-19 condition or 'Long Covid'[41] and affected patients have been referred to as 'long haulers'.[42] There is no universal case definition, but the CDC recently proposed defining late sequelae as sequelae that extend >4 weeks after initial infection.[43] The WHO has a clinical case definition of Long Covid:

> Post COVID-19 condition occurs in individuals with a history of probable or confirmed SARS CoV-2 infection, usually 3 months from the onset of COVID-19 with symptoms and that last for at least 2 months and cannot be explained by an alternative diagnosis.[44]

Symptoms maybe new onset following recovery from COVID-19 or persist from the initial illness. Symptoms may also fluctuate or relapse over time.

Common features of Long Covid are fatigue, joint pain, chest pain, palpitations, breathlessness, cognitive impairment and worsened quality of life.[45] A study from China has reported that most common symptoms were fatigue or muscle weakness and sleep difficulties.[46] Persistent symptoms after acute COVID-19 have also been reported in pregnant people[47] as well as children.[48]

Ongoing myocardial inflammation, lung fibrosis and dysfunction have been reported in various studies and are thought to be responsible for persistence of symptoms after discharge from hospital. Long Covid features are also seen in patients who were treated at the outdoor facilities. Neuropsychiatric symptoms have also been reported among patients who have recovered from acute COVID-19. High rates of anxiety and depression have been reported. Patients may continue to experience headaches, vision changes, hearing loss, loss of taste or smell, impaired mobility, numbness in extremities, tremors, myalgia, memory loss, cognitive impairment and mood changes even three months after diagnosis of COVID-19.[49]

Treatment of COVID-19

In the early phase of the pandemic there was little understanding of the disease-causing mechanism of SARS-CoV-2. As a result treatment options were limited to oxygen therapy, steroid (dexamethasone), remdesivir, hydroxychloroquine, ivermectin and mechanical ventilation for critically ill patients. Many of these medicines are no longer recommended for treatment. Intense research was carried out all over the world and as a result significant progress has been made both in understanding the disease process as well development of new treatment methods and vaccines at an unprecedented speed.

At present the available therapeutic options can be categorized into the following groups:[50]

- Antiviral drugs: molnupiravir, paxlovid (nirmatrelvir/ritonavir), remdesivir;
- Anti-SARS-COV-2 monoclonal antibodies:[51] sotrovimab, bamlanivimab/etesevimab, casirivimab/imdevimab;
- Anti-inflammatory agent: steroid (i.e., dexamethasone);
- Immunomodulators: baricitinib, tocilizumab.

The use of all these medicines depends upon severity of the disease and presence of certain patient-specific risk factors. In the initial phase of the disease antiviral medicines and antibody-based treatment is likely to be effective as the virus is in the phase of replication inside the body. In a later phase of the disease there is a hyperinflammatory stage where anti-inflammatory medicine and immunomodulators are going to help or a combination therapy may be successful.[52]

Mild disease

Patients suffering from mild disease should be isolated at home. Antipyretics are given to treat fever. Honey[53] and cough medicines are recommended to treat cough. Patients are advised to take oral fluids to prevent dehydration.

The WHO recommends sotrovimab or casirivimab/imdevimab for patients with non-severe disease who are at highest risk of hospitalization.[54] Casirivimab/imdevimab is not useful in Omicron infection. The National Institute for Health and Care Excellence in the UK and the National Institutes of Health in the USA recommend the use of monoclonal antibodies as well. The U.S. Food and Drug

Administration has also announced that as of April 5, 2022, sotrovimab is no longer authorized in the US, as the Omicron BA.2 subvariant accounts for >50% of cases.[55]

In the US, the National Institutes of Health[56] and in UK the National Institute for Health and Care Excellence[57] recommend antiviral remdesivir for patients suffering from mild to moderate disease. It should be started within seven days of disease onset.

WHO recommends molnupiravir, a new antiviral in adults with non-severe disease who are at highest risk of hospitalization (e.g., older age, immunosuppression and/or chronic diseases, unvaccinated for COVID-19). It should preferably be started within five days of disease onset and it is not recommended for children, pregnant or breastfeeding patients. Similarly, in the UK and USA, paxlovid (nirmatrelvir/ritonavir), a recently developed antiviral medicine, has been approved for the treatment of mild to moderate disease in adults who are at high risk of developing severe disease.[58]

Moderate disease

Patients with moderate COVID-19 illness should be hospitalized for close monitoring. All hospitalized patients should receive supportive care with adequate nutrition and appropriate rehydration and supplemental oxygen therapy must be initiated if SpO_2 is low and be maintained no higher than 96%.

Antibiotics are needed if there is suspicion of bacterial infection. Antipyretic paracetamol is recommended in these patients. These patients are at risk of developing blood clots; hence, prophylactic blood thinners are recommended. The National Institutes of Health has recommended antiviral remdesivir and anti-inflammatory dexamethasone to be considered for patients who are hospitalized and require supplemental oxygen. Recently a study published in reputed journal *NEJM* has shown that a three-day course of remdesivir resulted in an 87% lower risk of hospitalization or death among non-hospitalized patients who were at high risk for disease progression.[59] All these patients need psychological support.

Severe/critical disease

All these patients are hospitalized in an intensive care unit and put on oxygen therapy to maintain oxygen saturation. Pregnant women should be managed by a multidisciplinary team, including obstetric, perinatal, neonatal and intensive care specialists in a well-equipped centre.

Oxygen therapy is the mainstay of treatment. If saturation of oxygen is not maintained by giving oxygen through nasal prongs or a mask, the patient may require oxygen by high flow nasal cannula or noninvasive or invasive mechanical ventilation. Mechanical ventilation may be done in prone position to improve oxygenation. The National Institutes of Health (NIH) strongly recommends using dexamethasone in hospitalized patients who require oxygen via noninvasive or invasive ventilation. Combination therapy with dexamethasone plus remdesivir or baricitinib or tocilizumab in combination with dexamethasone alone is also recommended in hospitalized patients.[60] The WHO and National Institutes of Health strongly recommend the use of tocilizumab in serious/critical patients.[61]

These patients require anticoagulants and they should be started early. These patients often develop renal failure and may require dialysis. Besides this, the family members of the patients also need counselling to tide over the crisis.

Vaccine

Extraordinary efforts by clinical researchers at various institutes worldwide have resulted in the development of novel vaccines against SARS-CoV-2 at an unprecedented speed to contain this viral pandemic. Vaccination stimulates the immune system of the body and leads to production of neutralizing antibodies against SARS-CoV-2. According to the WHO COVID-19 dashboard,11,811,627,599 vaccine doses have been administered till May 23, 2022, throughout the world.[62]

The WHO has recommended the following vaccines for use globally.[63]

- mRNA vaccines: Pfizer/BioNTech, Moderna;
- Adenovirus vector vaccines: AstraZeneca, Janssen;
- Protein subunit vaccines: Novavax, Serum Institute of India;
- Inactivated virus vaccines: Sinopharm, Sinovac.

Pfizer/BioNTech vaccine

The Pfizer/BioNTech (BNT162b2) vaccine received U.S. Food and Drug Administration (FDA) approval on August 23, 2021, for individuals aged 16 years and older. Earlier the FDA issued a Emergency Use Authorization (EUA) on December 11, 2020, granting the use of the BNT162b2 vaccine to prevent COVID-19. It is given as an intramuscular injection in the muscle of the upper arm. Two doses are given in the primary series three to eight weeks apart. People aged 12 years and older who received a Pfizer-BioNTech primary series should get a booster. A second booster may be required for a selected group of people. In children between 5 and 11 years old, two doses of Pfizer-BioNTech are given three weeks apart.[64] Comirnaty is the brand name for the Pfizer-BioNTech COVID-19 vaccine.[65]

Vaccines reduce the risk of COVID-19, including the risk of severe illness and death among people who are fully vaccinated. It provides instructions the human body uses to build a harmless protein from the SARS-CoV-2 virus. This protein builds up an immune response that in turn protects the individual from getting sick with COVID-19.

Moderna vaccine

The Moderna vaccine (mRNA-1273) received an EUA from the FDA on December 18, 2020, granting the use of the mRNA-1273 vaccine to prevent COVID-19.

It received FDA approval on January 31, 2022, for individuals aged 18 years and older. Spikevax is the brand name for the Moderna COVID-19 vaccine. In the primary series two doses are given four to eight weeks apart. A booster dose is also recommended. It is given as an intramuscular injection in the muscle of the upper arm. Vaccines reduce the risk of COVID-19, including the risk of severe illness and death among people who are fully vaccinated and it acts in the same manner as the Pfizer/BioNTech vaccine.[66]

Janssen vaccine

The Janssen vaccine (JNJ-78436735) is a virus vector vaccine manufactured by Janssen Pharmaceuticals Companies of Johnson & Johnson. On February 27, 2021, the FDA issued an EUA for the Janssen COVID-19 vaccine for prevention of COVID-19 in adults aged 18 years and older.[67]

The Janssen vaccine is an adenovirus vector vaccine. This vector virus does not cause COVID-19. The vector virus when introduced builds up an immune response that protects the vaccinated person from falling sick due to COVID-19 in the future. After the production of the immune response the body discards all of the vaccine ingredients.

It is used as a single primary vaccination dose for individuals 18 years of age and older and as a single booster dose for individuals 18 years of age and older at least two months after completing primary vaccination with the vaccine. The Janssen COVID-19 vaccine is also authorized for use as a heterologous (or 'mix and match') single booster dose for individuals 18 years of age and older, following completion of primary vaccination with a different available COVID-19 vaccine.[68] People aged more than 18 years who are moderately or severely immunocompromised and who received the Janssen vaccine should get a second dose of an mRNA COVID-19 vaccine (Pfizer-BioNTech or Moderna).[69] It is given as an intramuscular injection in the muscle of the upper arm and is effective against COVID-19. However, in most situations, Pfizer-BioNTech or Moderna COVID-19 vaccines are preferred over the J&J/Janssen COVID-19 vaccine for primary and booster vaccination due to the risk of serious adverse events like blood clots with low platelet count.[70]

On May 5, 2022, the U.S. Food and Drug Administration has limited the authorized use of the Janssen COVID-19 vaccine to individuals 18 years of age and older for whom other authorized or approved COVID-19 vaccines are not accessible or clinically appropriate, and to individuals 18 years of age and older who elect to receive the Janssen COVID-19 vaccine because they would otherwise not receive a COVID-19 vaccine in view of the serious adverse event of blood clots with low platelet count.[71]

AstraZeneca vaccine

AstraZeneca/Oxford (ChAdOx1 nCoV-19/AZD1222) vaccine is a virus vector vaccine. It has been approved or granted EUA to prevent COVID-19 in many countries across the world but has not yet received an EUA or approval from the FDA for use in the US.[72] It is available by the brand name Vaxzevria and Covishield (manufactured by the Serum Institute in India and distributed to several countries under COVAX). On February 15, 2021, WHO granted EUA to this vaccine.

The vaccine has been recommended for all age groups 18 and older. Two doses of 0.5 ml each are given intramuscularly in the upper arm four to twelve weeks apart. The mechanism of actions is similar to the Janssen vaccine. It has also been approved for a booster dose after the primary series of vaccination. It is not yet authorized for use in children.

Novavax vaccine

Novavax (NVX-CoV2373) is a recombinant SARS-CoV-2 nanoparticle genetically engineered protein subunit vaccine. The WHO gave an EUA for Nuvaxovid (NVX-CoV2373) and Covovax (NVX-CoV2373) vaccine against COVID-19 on December 20, 2021, and December 17, 2021, respectively. The Novavax vaccine will be manufactured in Europe under the trade name Nuvaxovid and has been approved by the European Medicines Agency, and in India, the vaccine will be manufactured by the Serum Institute of India under the trade name Covovax and has been approved by the Drugs Controller General of India.[73]

The vaccine is not recommended for people younger than 18 years of age and pregnant women. However, it is recommended for breastfeeding mothers. It is a two-dose (0.5 ml) vaccine given intramuscularly. The two doses should be administered with an interval of three to four weeks.

In addition to the commonly used vaccines mentioned previously, several other vaccines have been developed indigenously including Covaxin (India) and Sputnik V (Russia) and have been approved or granted emergency use authorization to prevent COVID-19 in many countries around the world.[74] Broadly the vaccination program differs from one country to another. On August 11, 2020, Russia became the first country in the world to approve its own COVID-19 vaccine Sputnik V, developed by the Gamaleya National Center of Epidemiology and Microbiology (Moscow, Russia) for mass immunization.[75]

Vaccine protection and breakthrough infection

Protection starts seven to fourteen days after the primary series of vaccination. However, breakthrough infection after the primary series of vaccination has been reported. Breakthrough infections may be asymptomatic and there are reports of death as well.[76] Breakthrough infections have been reported with the Omicron variant, including people who received a booster dose. Vaccinated people are more likely to experience breakthrough infections more than three months after the second vaccine dose. However, prior infection with severe acute respiratory syndrome coronavirus 2 (SARS-CoV-2) may be associated with a lower risk for breakthrough infection.[77]

Additional dose of vaccine

In moderately to severely immunocompromised people aged ≥5 years, the WHO recommends that the primary vaccination series for all vaccines should be extended to include an additional dose. It should be given at least one month and within three months after the primary series (or at the earliest opportunity if more than three months have elapsed). In the UK, the Joint Committee on Vaccination and Immunisation (JCVI) recommends an additional dose in severely immunocompromised people aged ≥5 years, at least eight weeks after the second dose.[78]

Booster dose of vaccine

In the UK, the JCVI has recommended that all people aged 16 years or above should be offered a first booster dose at least three months after the completion of the primary course. Both the mRNA vaccines (Pfizer/BioNTech and Moderna) should be used with equal preference. A second booster dose is recommended in adults aged 75 years or older, for residents in care homes, and people 12 years or older who are immunosuppressed, around six months after the last dose, in March to May (Spring) 2022.[79]

In the US, the CDC recommends that all people aged 12 years or older should receive a first booster dose at least five months after completing their primary series of mRNA vaccine and at least two months after the Janssen vaccine. An mRNA vaccine is preferred over the Janssen vaccine for booster vaccination. A second booster dose of an mRNA vaccine is recommended at least four months after the first booster dose in people aged 12 years or older who are moderately or severely immunocompromised, and all adults aged 50 years or older.[80]

The WHO has recommended that the introduction of booster doses should be firmly evidence-driven and targeted to the population groups at highest risk of serious disease and those necessary to protect the health system.[81] A booster dose is recommended for the highest priority-use groups four to six months after the completion of the primary series. An ongoing trial has shown that a booster dose administered a median of 10.8 months after the second dose provided 95.3% efficacy compared with two doses during a median follow-up of 2.5 months.[82]

Heterologous vaccination

The WHO has recommended that homologous vaccination (the same vaccine for both the doses in the primary series) schedules should be considered standard practice. However, two heterologous doses of any authorized vaccine may be used to be a complete primary series.[83] Studies[84] have found that heterologous vaccination schedules[85] induce a robust immune response after a second dose of an mRNA vaccine in people primed with the AstraZeneca vaccine eight to twelve weeks earlier.[86]

Vaccine contraindications

Severe life threatening allergic reaction (anaphylaxis) and hypersensitivity to the vaccine[87] are absolute contraindication for vaccination.[88] People who have had thrombosis with thrombocytopenia (blood clot with low platelet) syndrome following the first dose of the AstraZeneca vaccine should not receive a second dose of the vaccine. Similarly if a person develops transverse myelitis (inflammation of spinal cord) after the first dose of the AstraZeneca vaccine he or she should not receive the second dose.[89] Delayed vaccination is recommended in people with current acute COVID-19 or any other acute febrile illness.[90]

Vaccine side effects

Side effects are common after COVID-19 vaccination. Almost all the organ systems may be adversely affected. However the common side effects observed are headache, joint pain, muscle pain, injection-site reactions, fatigue, fever, chills, nausea, vomiting, diarrhoea, syncope (after administration), rash and influenza-like reaction. Most of these side effects resolve within a day or two. Paracetamol is often required to treat them.

Some not-so-common but serious side effects may involve the cardiac and nervous system. There are reports of myocarditis (inflammation of the heart muscle) and pericarditis (inflammation of pericardium, the membrane covering the heart) especially after mRNA vaccination.[91] These side effects predominantly occur in adolescents and young adults, more often in males than in females, more often following the second dose and typically within three days after vaccination (up to 25 days).[92]

Guillain-Barre syndrome (inflammation of peripheral nerves leading to loss of muscle power) may occur following vaccination with adenovirus vector vaccines. Rarely cases have also been reported with mRNA vaccine.[93] Transverse myelitis may occur rarely following vaccination.[94] Features include muscle weakness, localized or radiating back pain, bladder and bowel symptoms, and changes in sensation.[95]

Severe allergic reactions, including anaphylaxis (severe potential life threatening allergy) may occur after vaccination. Thrombocytopenia (low platelet count) has been reported as a side effect of vaccination. Similarly, vaccine-induced immune thrombocytopenia and thrombosis (VITT) have also been reported. In this condition there are blood clots with low platelet count.[96]

Reports of death after COVID-19 vaccination are rare. More than 584 million doses of COVID-19 vaccines were administered in the United States from December 14, 2020, through May 23, 2022. During this time, 14,778 preliminary reports of death (0.0025%) were reported among people who received a COVID-19 vaccine.[97]

Vaccination of pregnancy and breastfeeding

The WHO, CDC, the JCVI and the Royal College of Obstetricians and Gynaecologists in the UK and the American College of Obstetricians and Gynecologists recommend that all pregnant women, or women who are thinking about or trying to become pregnant, and breastfeeding women should be vaccinated.[98]

Vaccine efficacy

Overall initial reported vaccine efficacy for preventing symptomatic infection was reported as 95% (Pfizer/BioNTech), 94.1% (Moderna), 74% (AstraZeneca), and 66.9% (Janssen).[99] However, vaccine efficacy depends on the SARS-CoV-2 variant. Efficacy is highest for the Alpha variant, with lower efficacy reported for Beta and Gamma variants. Vaccine efficacy (in terms of testing positive and onwards transmission) for the Delta variant appears to be reduced relative to other variants.[100]

The study published in the journal *NEJM*[101] has summarized the effectiveness of vaccines against COVID-19 and states that,

> Coronavirus disease 2019 (Covid-19) vaccines are highly effective against symptomatic disease and, more so, against severe disease and fatal outcomes caused by the original strain of SARS-CoV-2 as well as the alpha (B.1.1.7) variant that predominated in early 2021. Modest reductions in vaccine effectiveness against infection and mild disease have been observed with the beta (B.1.351) and delta variants, although effectiveness against severe disease has remained high for at least 6 months after primary immunization with two Covid-19 vaccine doses. Waning of protection has been observed with time since vaccination, especially with the delta variant, which is able to at least partially evade natural and vaccine-induced immunity. However, third (booster) doses provide a rapid and substantial increase in protection against both mild and severe disease.

This study further concluded that primary immunization with two doses of AstraZeneca or Pfizer vaccine provided limited protection against symptomatic disease caused by the Omicron variant. A Pfizer or Moderna booster after either the AstraZeneca or Pfizer primary series of vaccination substantially increased protection, but that protection waned over time.[102]

Vaccine hesitancy

The WHO defines vaccine hesitancy as a 'delay in acceptance or refusal of vaccines despite availability of vaccination services'.[103] There has been a steady rise of COVID-19 cases and deaths and at the same time there is progressive increase in COVID-19 vaccine supplies, but the hesitancy and refusal to be vaccinated has turned out to be a major problem in many parts of the world.

Reasons for vaccine hesitancy fit into three categories: lack of *confidence* (in effectiveness, safety, the system or policy makers), *complacency* and lack of *convenience*.[104] Vaccine hesitancy allows the coronavirus to continue spreading in the community and leads to the emergence of new variants. The trust of people on the safety of vaccines is very low as the vaccines against COVID-19 have been prepared in an emergency situation at an unprecedented speed in various institutes of the world.[105] In May 2020, about 25% of people in five surveys stated that they would refuse a future vaccine if it would have been available, mainly due to safety concerns.[106]

The spread of misinformation and disinformation on social media and through other channels has affected COVID-19 vaccine confidence. Misinformation is false information shared by people who do not intend to mislead others. Disinformation is false information deliberately created and disseminated with malicious intent. Most misinformation and disinformation that has circulated about COVID-19 vaccines has focused on vaccine development, safety and effectiveness, as well as COVID-19 denialism resulting in vaccine hesitancy throughout the world.[107]

A study published in the reputed journal *Public Health* has observed that the most common reasons to refuse vaccination were being against vaccines in general; having concerns about safety, thinking that a vaccine produced in a rush is too dangerous; considering the vaccine useless because of the harmless nature of COVID-19; having a general lack of trust, doubts about the efficiency and provenience of the vaccine; and possessing a belief to be already immunized.[108]

The CDC has highlighted the following steps to build trust on COVID-19 vaccines:[109]

- There should be transparent communication about the process of authorizing, approving, making recommendations for, monitoring the safety of, distributing, allocating and administering of COVID-19 vaccines including data handling;
- Regular updates should be provided on safety, benefits, side effects and effectiveness of COVID-19 vaccines;
- Every effort should be made at appropriate levels to help stop the spread and harm of misinformation via social media platforms, partners and trusted messengers.

Until and unless people trust the vaccines, vaccine hesitancy will not disappear. Widespread vaccination and uniform global coverage are probably the only way we can check the spread of the COVID-19 pandemic and breathe a sigh of relief.

In future science will definitely find out better ways of managing the COVID-19 pandemic in the form of new medicines and vaccines. However, every effort will fail if robust global healthcare architecture is not created whereby the gain obtained is transferred equally and effectively universally. We have to remember that we are safe only when everyone is safe.

Notes

1 Septic shock is a serious condition in which there is infection throughout the body leading to drop in blood pressure.
2 Cascella M, et al., Features, evaluation, and treatment of coronavirus (COVID-19), available at www.ncbi.nlm.nih.gov/books/NBK554776/, accessed on April 28, 2022.

3 Islam MA, et al., Prevalence and characteristics of fever in adult and paediatric patients with coronavirus disease 2019 (COVID-19): A systematic review and meta-analysis of 17515 patients, *PLoS One*, 2021; 16(4): e0249788.

4 Song WJ, et al., Confronting COVID-19-associated cough and the post-COVID syndrome: Role of viral neurotropism, neuroinflammation, and neuroimmune responses, *Lancet Respir Med*, 2021; 9(5): 533–544.

5 Hariyanto TI, et al., Anosmia/hyposmia is a good predictor of coronavirus disease 2019 (COVID-19) infection: A meta-analysis, *Int Arch Otorhinolaryngol*, 2021; 25(1): e170–e174.

6 Coronavirus Disease 2019 (COVID-19), *BMJ Best Practice*, available at https://best-practice.bmj.com/topics/en-gb/3000201/history-exam, accessed on April 28, 2022.

7 Chen A, et al., Are gastrointestinal symptoms specific for coronavirus infection? A prospective case-control study from the United States, *Gastroenterology*, 2020; 159(3): 1161–1163.

8 Bolia R, et al., Gastrointestinal manifestations of pediatric coronavirus disease and their relationship with a severe clinical course: A systematic review and meta-analysis, *J Trop Pediatr*, 2021; 67(2): fmab051.

9 Cénat JM, et al., Prevalence of symptoms of depression, anxiety, insomnia, posttraumatic stress disorder, and psychological distress among populations affected by the COVID-19 pandemic: A systematic review and meta-analysis, *Psychiatry Res*, 2020; 295: 113599.

10 Nasiri N, et al., Ocular manifestations of COVID-19: A systematic review and meta-analysis, *J Ophthalmic Vis Res*, 2021; 16(1): 103–112.

11 Visconti A, et al., Diagnostic value of cutaneous manifestation of SARS-CoV-2 infection, *Br J Dermatol*, 2021; 184(5): 880–887.

12 Daneshgaran G, et al., Cutaneous manifestations of COVID-19: An evidence-based review, *Am J Clin Dermatol*, 2020; 21(5): 627–639.

13 Martinez-Rojas MA, et al., Is the kidney a target of SARS-CoV-2? *Am J Physiol Renal Physiol*, 2020; 318(6): F1454–F1462.

14 Inflammation of the muscle of heart.

15 Hessami A, et al., Cardiovascular diseases burden in COVID-19: Systematic review and meta-analysis, *Am J Emerg Med*, 2021; 46: 382–391.

16 Kim D, et al., Rates of co-infection between SARS-CoV-2 and other respiratory pathogens, *JAMA*, 2020; 323(20): 2085–2086, available at www.ncbi.nlm.nih.gov/pub-med/32293646, accessed on April 28, 2022.

17 *Mucormycosis* is a serious but rare fungal infection caused by a group of moulds called mucormycetes.

18 Kumar Singh A, et al., Mucormycosis in COVID-19: A systematic review of cases reported worldwide and in India, *Diabetes Metab Syndr*, 2021; 15(4): 102146, available at www.ncbi.nlm.nih.gov/pmc/articles/PMC8137376/, accessed on April 28, 2022.

19 Nucleic acid amplification test.

20 Covid-19 treatment guidelines, NIH, available at www.covid19treatmentguidelines.nih.gov/overview/clinical-spectrum/, accessed on April 28, 2022.

21 Oxygen saturation.

22 Ferrando C, et al., COVID-19 Spanish ICU network. Clinical features, ventilatory management, and outcome of ARDS caused by COVID-19 are similar to other causes of ARDS. *Intensive Care Med*, 2020; 46(12): 2200–2211.

23 COVID-19 (coronavirus disease): People with certain medical conditions, 2020, CDC, available at www.cdc.gov/coronavirus/2019-ncov/need-extra-precautions/people-with-medical-conditions.html, accessed on April 28, 2022.

24 Coronavirus (COVID-19) infection in pregnancy, RCOG, available at chrome-extension://efaidnbmnnnibpcajpcglclefindmkaj/www.rcog.org.uk/media/xsubnsma/2022-03-07-coronavirus-covid-19-infection-in-pregnancy-v15.pdf, accessed on April 28, 2022.

25 Coronavirus disease (COVID-19): Pregnancy, childbirth and the postnatal period, WHO, available at www.who.int/news-room/questions-and-answers/item/coronavirus-disease-covid-19-pregnancy-and-childbirth, accessed on April 28, 2022.

26 Ibid.

27 Ibid.

28 Coronavirus (COVID-19) infection in pregnancy, RCOG, available at https://www.rcog.org.uk/media/xsubnsma/2022-03-07-coronavirus-covid-19-infection-in-pregnancy-v15.pdf, accessed on April 28, 2022.

29 Society for Maternal Fetal Medicine, Management considerations for pregnant patients with COVID-19, 2020, available at https://s3.amazonaws.com/cdn.smfm.org/media/2336/SMFM_COVID_Management_of_COVID_pos_preg_patients_4-30-20_final.pdf, accessed on April 28, 2022.

30 COVID-19, Pregnancy, childbirth, and breastfeeding: Answers from Ob-Gyns, ACOG, available at www.acog.org/womens-health/faqs/coronavirus-covid-19-pregnancy-and-breastfeeding, accessed on April 29, 2022.

31 Pregnant and recently pregnant people at increased risk for severe illness from COVID-19, CDC, available at www.cdc.gov/coronavirus/2019-ncov/need-extra-precautions/pregnant-people.html, accessed on April 29, 2022.

32 Hoang A, et al., COVID-19 in 7780 pediatric patients: A systematic review, *E Clinical Med*, 2020; 24(100433). https://doi.org/10.1016/j.eclinm.2020.100433.

33 Ravindra K, Consideration for the asymptomatic transmission of COVID-19: Systematic review and meta analysis, *MedRxiv*, 2020: 2020, available at https://www.medrxiv.org/content/10.1101/2020.10.06.20207597v1, accessed on July 30, 2022.

34 Jiang L, et al., COVID-19 and multisystem inflammatory syndrome in children and adolescents, *Lancet Infect Dis*, 2020; 20(11): e276–e288, https://doi.org/10.1016/S1473-3099(20)30651-4

35 For parents: Multisystem inflammatory syndrome in children (MIS-C) associated with COVID-19, CDC, available at www.cdc.gov/mis/mis-c.html, accessed on April 29, 2022.

36 COVID-19 disease in children and adolescents, WHO Scientific Brief, available at https://apps.who.int/iris/rest/bitstreams/1375120/retrieve, accessed on April 29, 2022.

37 Viner RM, et al., Susceptibility to SARS-CoV-2 infection among children and adolescents compared with adults: A systematic review and meta-analysis, *JAMA Pediatr*, 2021; 175: 143–156, https://doi.org/10.1001/jamapediatrics; 2020.4573

38 Götzinger F, et al., COVID-19 in children and adolescents in Europe: A multinational, multicentre cohort study, *Lancet Child Adolesc Health*, 2020; 4: 653–661.

39 Public Health England, PHE weekly national influenza and COVID-19 report: Week 1 report (up to week 53 data), available at https://assets.publishing.service.gov.uk/government/uploads/system/uploads/attachment_data/file/954733/Weekly_Influenza_and_COVID19_report_data_W1_V2.xlsx, accessed on May 6, 2022.

40 Olabi B, et al., Population perspective comparing COVID-19 to all and common causes of death during the first wave of the pandemic in seven European countries, *Public Health in Pract (Oxf)*, 2021; 2100077.

41 Coronavirus disease (COVID-19): Post COVID-19 condition, WHO, available at www.who.int/news-room/questions-and-answers/item/coronavirus-disease-(covid-19)-post-covid-19-condition#:~:text=Post%20COVID%2D19%20condition%2C%20also,as%20%E2%80%9Clong%2Dhaulers.%E2%80%9D, accessed on April 29, 2022.

42 Clinical spectrum of SARS-CoV-2 infection, NIH, available at www.covid19treatmentguidelines.nih.gov/overview/clinical-spectrum/, accessed on April 29, 2022.

43 Datta SD, et al., A proposed framework and timeline of the spectrum of disease due to SARS-CoV-2 infection: Illness beyond acute infection and public health implications, *JAMA*, 2020; 324(22): 2251–2252, available at www.ncbi.nlm.nih.gov/pubmed/33206133, accessed on May 6, 2022.

44 A clinical case definition of post COVID-19 condition by a Delphi consensus, October 6, 2021, WHO, available at www.who.int/publications/i/item/WHO-2019-nCoV-Post_COVID-19_condition-Clinical_case_definition-2021.1, accessed on April 29, 2022.

45 Halpin SJ, et al., Postdischarge symptoms and rehabilitation needs in survivors of COVID-19 infection: A cross-sectional evaluation, *J Med Virol*, 2021; 93(2): 1013–1022, available at www.ncbi.nlm.nih.gov/pubmed/32729939, accessed on May 6, 2022.

46 Huang C, et al., 6-month consequences of COVID-19 in patients discharged from hospital: A cohort study, *Lancet*, 2021; 397(10270): 220–232, available at www.ncbi.nlm.nih.gov/pubmed/33428867, accessed on May 6, 2022.

47 Afshar Y, et al., Clinical presentation of coronavirus disease 2019 (COVID-19) in pregnant and recently pregnant people, *Obstet Gynecol*, 2020; 136(6): 1117–1125, available at www.ncbi.nlm.nih.gov/pubmed/33027186, accessed on May 6, 2022.

48 Stephenson T, et al., Long COVID – the physical and mental health of children and non-hospitalised young people 3 months after SARS-CoV-2 infection; a national matched cohort study (The CLoCk) Study, available at www.researchsquare.com/article/rs-798316/v1, accessed on April 29, 2022.

49 Clinical spectrum of SARS-CoV-2 infection, NIH, available at www.covid19treatmentguidelines.nih.gov/overview/clinical-spectrum/, accessed on April 29, 2022.

50 Coopersmith CM, et al., The surviving sepsis campaign: Research priorities for coronavirus disease 2019 in critical illness, *Crit Care Med*, 2021; 49(4): 598–622.

51 A *monoclonal antibody* is a molecule developed in a *laboratory* that is designed to mimic or enhance the body's natural immune system response against microorganisms. They have the ability to resist the virus.

52 Cascella M, et al., Features, evaluation, and treatment of coronavirus (COVID-19), available at www.ncbi.nlm.nih.gov/books/NBK554776/, accessed on April 28, 2022.

53 Abuelgasim H, et al., Effectiveness of honey for symptomatic relief in upper respiratory tract infections: A systematic review and meta-analysis, *BMJ Evid Based Med*, 2021; 26(2): 57–64.

54 Coronavirus Disease 2019 (COVID-19), *BMJ Best Practice*, available at https://bestpractice.bmj.com/topics/en-gb/3000201/history-exam, accessed on April 28, 2022.

55 Ibid.

56 National Institutes of Health, Coronavirus disease 2019 (COVID-19) treatment guidelines, 2022, available at www.covid19treatmentguidelines.nih.gov/about-the-guidelines/whats-new/, accessed on May 6, 2022.

57 National Institute for Health and Care Excellence, COVID-19 rapid guideline: Managing COVID-19, May 2022, available at www.nice.org.uk/guidance/ng191/resources/managing-covid-19-treatments-may-2022-v24.0-pdf-11070542125, accessed on May 6, 2022.

58 Coronavirus Disease 2019 (COVID-19), *BMJ Best Practice*, available at https://bestpractice.bmj.com/topics/en-gb/3000201/history-exam, accessed on April 28, 2022.

59 Gottlieb RL, et al., Early remdesivir to prevent progression to severe Covid-19 in outpatients, *N Engl J Med*, 2022; 386(4): 305–315.

60 Cascella M, et al., Features, evaluation, and treatment of coronavirus (COVID-19), available at www.ncbi.nlm.nih.gov/books/NBK554776/, accessed on April 28, 2022.

61 Coronavirus Disease 2019 (COVID-19), *BMJ Best Practice*, available at https://bestpractice.bmj.com/topics/en-gb/3000201/history-exam, accessed on April 28, 2022.

62 WHO, available at https://covid19.who.int/, accessed on May 28, 2022.

63 Coronavirus Disease 2019 (COVID-19), *BMJ Best Practice*, available at https://bestpractice.bmj.com/topics/en-gb/3000201/history-exam, accessed on April 28, 2022.

64 Stay up to date with your COVID-19 vaccines, CDC, available at www.cdc.gov/coronavirus/2019-ncov/vaccines/stay-up-to-date.html, accessed on May 1, 2022.

65 Pfizer-BioNTech COVID-19 vaccine (also known as COMIRNATY): Overview and safety, CDC, available at www.cdc.gov/coronavirus/2019-ncov/vaccines/different-vaccines/Pfizer-BioNTech.html, accessed on May 1, 2022.

66 Ibid.

67 ACIP evidence to recommendations for use of Janssen COVID-19 vaccine under an emergency use authorization, CDC, available at www.cdc.gov/vaccines/acip/recs/grade/covid-19-janssen-etr.html, accessed on May 1, 2022.

68 Janssen COVID-19 vaccine, US FDA, available at www.fda.gov/emergency-preparedness-and-response/coronavirus-disease-2019-covid-19/janssen-covid-19-vaccine, accessed on May 1, 2022.

69 Johnson & Johnson's Janssen COVID-19 vaccine: Overview and safety, CDC, available at www.cdc.gov/coronavirus/2019-ncov/vaccines/different-vaccines/janssen.html, accessed on May 1, 2022.

70 Ibid.

71 Coronavirus (COVID-19) update: FDA limits use of Janssen COVID-19 vaccine to certain individuals, FDA, available at www.fda.gov/news-events/press-announcements/coronavirus-covid-19-update-fda-limits-use-janssen-covid-19-vaccine-certain-individuals, accessed on May 8, 2022.

72 Cascella M, et al., Features, evaluation, and treatment of coronavirus (COVID-19), available at www.ncbi.nlm.nih.gov/books/NBK554776/, accessed on April 28, 2022.

73 The Novavax vaccine against COVID-19: What you need to know, WHO, available at www.who.int/news-room/feature-stories/detail/the-novavax-vaccine-against-covid-19-what-you-need-to-know, accessed on May 1, 2022.

74 Cascella M, et al., Features, evaluation, and treatment of coronavirus (COVID-19), available at www.ncbi.nlm.nih.gov/books/NBK554776/, accessed on April 28, 2022.

75 Khan Burki Y, The Russian vaccine for COVID-19, *The Lancet Resp Med*, available at www.thelancet.com/journals/lanres/article/PIIS2213-2600(20)30402-1/fulltext, accessed on May 8, 2022.

76 Zhang M, et al., A systematic review of vaccine breakthrough infections by SARS-CoV-2 Delta variant, *Int J Biol Sci*, 2022; 18(2): 889–900, available at www.ncbi.nlm.nih.gov/pmc/articles/PMC8741840/, accessed on May 6, 2022.

77 Coronavirus Disease 2019 (COVID-19), *BMJ Best Practice*, available at https://bestpractice.bmj.com/topics/en-gb/3000201/history-exam, accessed on April 28, 2022.

78 Ibid.

79 UK Health Security Agency, JCVI advises a spring COVID-19 vaccine dose for the most vulnerable, 2022.

80 Coronavirus Disease 2019 (COVID-19), *BMJ Best Practice*, available at https://bestpractice.bmj.com/topics/en-gb/3000201/history-exam, accessed on April 28, 2022.

81 World Health Organization, Interim statement on booster doses for COVID-19 vaccination, 2021, available at www.who.int/news/item/22-12-2021-interim-statement-on-booster-doses-for-covid-19-vaccination–update-22-december-2021, accessed on May 6, 2022.

82 Moreira ED Jr, et al., Safety and efficacy of a third dose of BNT162b2 Covid-19 vaccine, *N Engl J Med*, 2022; 386: 1910–1921, https://doi.org/10.1056/NEJMoa2200674

83 Coronavirus Disease 2019 (COVID-19), *BMJ Best Practice*, available at https://bestpractice.bmj.com/topics/en-gb/3000201/history-exam, accessed on April 28, 2022.

84 Borobia AM, et al., Immunogenicity and reactogenicity of BNT162b2 booster in ChAdOx1-S-primed participants (CombiVacS): A multicentre, open-label, randomised, controlled, phase 2 trial, *Lancet*, 2021; 398(10295): 121–130.

85 Normark J, et al., Heterologous ChAdOx1 nCoV-19 and mRNA-1273 vaccination, *N Engl J Med*, 2021; 385(11): 1049–1051.

86 Hillus D, et al., Safety, reactogenicity, and immunogenicity of homologous and heterologous prime-boost immunisation with ChAdOx1 nCoV-19 and BNT162b2: A prospective cohort study, *Lancet Respir Med*, 2021; 9(11): 1255–1265.

87 World Health Organization (WHO), Interim recommendations for use of the ChAdOx1-S [recombinant] vaccine against COVID-19 (AstraZeneca COVID-19 vaccine AZD1222 Vaxzevria™, SII COVISHIELD™), 2022, available at www.who.int/publications/i/item/WHO-2019-nCoV-vaccines-SAGE_recommendation-AZD1222-2021.1, accessed on May 6, 2022.

88 World Health Organization (WHO), Interim recommendations for use of the Pfizer – BioNTech COVID-19 vaccine, BNT162b2, under emergency use listing, 2022, available at www.who.int/publications/i/item/WHO-2019-nCoV-vaccines-SAGE_recommendation-BNT162b2-2021.1, accessed on May 6, 2022.

89 Coronavirus Disease 2019 (COVID-19), *BMJ Best Practice*, available at https://bestpractice.bmj.com/topics/en-gb/3000201/history-exam, accessed on April 28, 2022.

90 Ibid.
91 Truong DT, et al., Clinically suspected myocarditis temporally related to COVID-19 vaccination in adolescents and young adults, *Circulation*, 2022; 145(5): 345–356.
92 Matta A, et al., Clinical presentation and outcomes of myocarditis post mRNA vaccination: A meta-analysis and systematic review, *Cureus*, 2021; 13(11): e19240.
93 Shafiq A, et al., Neurological immune-related adverse events post-COVID-19 vaccination: A systematic review, *J Clin Pharmacol*, 2022; 62(3): 291–303.
94 Maroufi SF, et al., Longitudinally extensive transverse myelitis after Covid-19 vaccination: Case report and review of literature, *Hum Vaccin Immunother*, 2022; 18(1): 2040239-1–2040239-4.
95 Coronavirus Disease 2019 (COVID-19), *BMJ Best Practice*, available at https://best-practice.bmj.com/topics/en-gb/3000201/history-exam, accessed on April 28, 2022.
96 Ibid.
97 Selected adverse events reported after COVID-19 vaccination, CDC, available at www.cdc.gov/coronavirus/2019-ncov/vaccines/safety/adverse-events.html, accessed on May 28, 2022.
98 Ibid.
99 Ibid.
100 Ibid.
101 Andrews N, Covid-19 vaccine effectiveness against the Omicron (B.1.1.529) variant, *NEJM*, available at www.nejm.org/doi/full/10.1056/NEJMoa2119451#:~:text=After%20a%20BNT162b2%20primary%20course,at%205%20to%209%20weeks, accessed on April 24, 2022.
102 Ibid.
103 MacDonald NE, SAGE Working Group on vaccine hesitancy vaccine hesitancy: Definition, scope and determinants, *Vaccine*, 2015; 33(34): 4161–4164, https://doi.org/10.1016/j.vaccine.2015.04.036
104 Shixin (Cindy) Shen and Vinita Dubey, Addressing vaccine hesitancy, *Can Fam Physician*, 2019; 65(3): 175–181.
105 Chou WS and Budenz A, Considering emotion in COVID-19 vaccine communication: Addressing vaccine hesitancy and fostering vaccine confidence, *Health Commun*, 2020; 35(14): 1718–1722.
106 Group C, A future vaccination campaign against COVID-19 at risk of vaccine hesitancy and politicisation, *Lancet Infect Dis*, 2020; 20(7): 769–770.
107 How to address COVID-19 vaccine misinformation, CDC, available at www.cdc.gov/vaccines/covid-19/health-departments/addressing-vaccine-misinformation.html, accessed on May 3, 2022.
108 Troiano G and Nardi A, Vaccine hesitancy in the era of COVID-19, *Public Health*, 2021; 194: 245–251, available at www.ncbi.nlm.nih.gov/pmc/articles/PMC7931735/#bib13, accessed on May 3, 2022.
109 Vaccinate with confidence COVID-19 vaccines strategy for adults, CDC, available at www.cdc.gov/vaccines/covid-19/vaccinate-with-confidence/strategy.html, accessed on May 3, 2022.

AFTERWORD TO THE FIRST EDITION

The COVID-19 pandemic caused by SARS-CoV-2, the seventh coronavirus to affect human beings, has been rapidly evolving throughout the world.[1] This has led to corresponding dynamic changes in almost all aspects of the pandemic, necessitating the inclusion of new information as this work goes to press.

Worldwide spread

According to WHO situation report 151, dated June 19, 2020, the total number of confirmed cases in the world has gone up to 8,385,440 and the number of deaths has climbed to 450,686.[2] The composition of the top ten worst-affected countries has also changed, as shown in Table 1.

New epicentre

According to the WHO, the Americas[3] have emerged as the new epicentre of the coronavirus pandemic, because a US study has forecasted deaths surging in Brazil and other Latin American countries through August 2020. Carissa Etienne, WHO director for the Americas and head of the Pan American Health Organization (PAHO), has expressed concern over the accelerating outbreaks in Peru, Chile, El Salvador, Guatemala and Nicaragua. Latin America has already passed Europe and the United States in daily infections.

A University of Washington study has projected that Brazil's total death toll could rise to 125,000 by early August.[4] Brazil's most populous state, Sao Paulo, has become the epicentre of the pandemic in Brazil, which reported a record number of COVID-19 deaths, accounting for a quarter of the country's total fatalities.[5] Brazil now has the second highest death toll (only after the United States), where the handling of the pandemic has turned highly political, resulting in difficulties in controlling the

TABLE 1 Top Ten Worst-Affected Countries in the World

Name of the country	Number of confirmed cases	Number of deaths
USA	2,149,166	117,472
Brazil	955,377	46,510
Russia	569,063	7841
India	380,532	12,573
UK	300,473	42,288
Spain	245,268	27,136
Peru	240,908	7257
Italy	238,159	34,514
Chile	225,103	3841
Iran	197,647	9272

Source: WHO.

pandemic.[6] In Brazil, two health ministers, both doctors, have been removed and replaced by Eduardo Pazuello, an interim minister without any medical background. The overall situation in Latin America, and Brazil in particular, is extremely worrisome, and the situation seems to be completely out of control. There is no sign of the pandemic stabilizing, and even the peak seems to be far off. It seems the pandemic will definitely continue even in the early part of 2021, with significant casualties.

China: the second wave

In the middle of May 2020, Chinese health authorities started testing millions of people in Wuhan, after the city reported a small cluster of infections.[7]

Chinese authorities sealed off the northeastern city of Shulan after an outbreak of coronavirus and imposed measures similar to those used in Wuhan.[8] The northeast of China, which borders Russia and North Korea, has emerged as a new hotspot. These cases are thought to be due to cross-border travel.

A new outbreak of great concern has started in Beijing.[9] The city has reported almost 200 new cases of coronavirus which are linked to the Xinfadi food market, located in the Southern Fengtai district. It has been shut down, along with two others, and at least 33 neighbourhoods have been put under varying levels of lockdown. Schools and other venues of sport and entertainment have been closed.[10]

China seemed to have controlled the pandemic after imposing the strictest possible lockdown measures in various cities. However, the recent spurt of cases in various places, especially in Beijing, points towards a second wave of the pandemic in China, which may continue for several months or even longer.

The United States

The pandemic is still spreading across various parts of the United States: Arizona, Florida, Oklahoma, Oregon and Texas all reported their highest numbers of new cases as they continued to reopen their economies.[11] Similarly, officials in New York and Houston are considering another lockdown to prevent an increase in new cases.[12] The number of cases in New York has declined, but a week of protests on the streets in the wake of the death of George Floyd in Minneapolis could bring a new wave.[13] Memorial Day weekend gatherings in late May are also thought to be responsible for the increase in the number of cases.

New cases have declined in places such as Illinois, Michigan and Washington, which were affected badly earlier.[14] However, the second wave of the pandemic affecting states such as Arizona, Arkansas, California, Florida, North Carolina, Oregon, South Carolina, Texas and Utah is of great concern. The CDC has projected up to 140,000 deaths in the United States by early July.[15] The pandemic in a large country like the United States is not likely to end by the end of 2020, and there is likely to be a significant increase in the number of cases and in the number of deaths.

Europe

The pandemic seems to have stabilized in the badly affected countries of Europe. Across Europe there is gradual and cautious lifting of lockdown measures. In France, the number of new cases continued to drop even as most of the country reopened for business after a strict eight-week lockdown, with the number of additional deaths down to its lowest figure since March.[16] In Italy, the situation is similar to France. However, Italian Prime Minister Giuseppe Conte was questioned by prosecutors about the country's response to the pandemic, which has killed more than 34,000 people.[17] The prosecutors from Bergamo are looking into why badly affected small towns around the city were not locked down earlier in the outbreak, in spite of rising number of cases. The UK has recorded its lowest daily rise in the number of coronavirus deaths since before lockdown started on March 23, 2020.[18] The pandemic seems to have passed its peak in the UK and is now on the decline. The situation in Germany is similar to that of the UK. However, gatherings of extended families in the northwestern German city of Göttingen have resulted in a local spike of the coronavirus.[19] In Belgium and Sweden, the pandemic is on the decline. These two countries have also passed the peak of the pandemic.[20] Although the pandemic is on the decline in Western Europe, this does not call for complacency, as WHO situation report 151, dated June 19, 2020, has informed that the UK has reported 181 new deaths and Italy has reported 66 new deaths in the past 24 hours. Similarly, France has also reported 28 new deaths.[21]

In Russia, the pandemic remains ongoing, and it is now the third worst-affected country in the world after the United States and Brazil. However, its official death toll is much lower than the figures seen in other countries with serious outbreaks.[22] Although the Kremlin has denied any problem with its official data, the WHO expressed concern. Moscow's health department raised its death toll for May, citing changes in the way it determines the cause of death for patients who have other health problems.[23] It had earlier more than doubled the official death toll from COVID-19 in the Russian capital for the month of April as well.[24] Russia reported 181 new deaths on June 19, 2020.[25] The end of the pandemic is not yet in sight in Russia and is projected to continue until the end of 2020.

The pandemic is still continuing in various other parts of Europe:[26] Turkey has reported 1304 new cases and 21 deaths; and Poland has reported 314 new cases and 30 deaths. Efforts must be made to control the surge of infections in various parts of Europe, and at the same time, the countries where the epidemic has passed the peak should not lower their guards, as a coronavirus outbreak may happen suddenly and spread beyond control again.

Asia, Africa and the rest of the world

India is now the worst-affected country in Asia and is only behind the United States, Brazil and Russia in terms of confirmed cases. With rapidly improving case detection facilities and by easing lockdown measures, the number of cases and deaths has started climbing rapidly. A recent study projected that the peak of the COVID-19 pandemic in India was delayed by the eight-week lockdown, along with strengthened public health measures, and it may now arrive around mid November.[27] Maharashtra, Tamil Nadu, Delhi and Gujarat are the worst-affected states, and Delhi overtook Maharashtra as regards coronavirus positivity rates, reaching 30% in the first two weeks of June 2020.[28]

There is stress in healthcare facilities due to increases in the number of cases, and the Indian government is trying to increase bed capacity by converting railway coaches, stadiums, banquet halls and hotels into healthcare facilities to treat COVID-19 patients.[29] Numerous frontline workers like doctors, nurses, teachers and police personnel have been affected by COVID-19. Mr Satyendar Jain, the health minister of Delhi, tested positive and has been admitted to a hospital.[30]

The overall situation is grim in India. India is a vast country with a huge population. If adequately tested, the number of cases may surpass the United States in the near future. The pandemic will peak in the densely populated cities in the beginning, which is most likely to happen in summer 2020, after which it is expected to spread in villages. Neither the peak nor the end of the pandemic is in sight at present. However, the pandemic may stabilize by the end of 2020, and the decline may start from the first half of 2021. India must be well prepared to tackle the situation that is going to unfold in the coming months.

In Iran, there has been a rapid surge in cases recently, sparking fears of a second wave of the pandemic.[31] New infections have been averaging more than 3000 a day in the first week of June. The worst-affected provinces are Khuzestan, Hormozgan, Kurdistan and Kermanshah.

Pakistan has also seen its number of infections and deaths rise in recent weeks, and the healthcare system is under strain.[32] Similarly, cases are also rising in Saudi Arabia and Qatar. The rise in cases and deaths in India, Iran, Pakistan and Saudi Arabia may turn this region into the next epicentre of the pandemic after the Americas.

After Africa hit 200,000 cases, the WHO warned that the pandemic is accelerating there.[33] South Africa is the worst-affected country on the continent, with more than a quarter of all its infections. Egypt has also been badly affected. The testing rate in Africa is very low, which makes it difficult to predict its future trend. However, the continent has the potential to turn into the new epicentre of the pandemic after the Americas and Asia.

New Zealand has lifted almost all of its coronavirus restrictions after reporting no active cases in the country.[34] New Zealand has reported no new COVID-19 cases for more than two weeks. However, recently a few cases have emerged that shattered the hope of eliminating the virus from New Zealand.[35]

Global mortality

The global burden from COVID-19 is assessed by the number of deaths occurring in all affected countries. However, due to various methods of testing, detecting cases and announcing death counts, country's respective numbers are difficult to compare. However, case fatality ratio (death per number of case) and death per population are two widely used methods. The mortality data compiled by Johns Hopkins University are represented in Tables 2 and 3 for the top ten worst-affected countries.[36]

In terms of the number of cases, India is the fourth worst-affected country, but in terms of mortality, it does not appear among the top ten worst-affected countries. The case fatality ratio in the country is 3.3%, and the deaths per 100,000 population ratio is 0.96%.[37]

TABLE 2 Top Ten Countries with the Worst Case Fatality Ratios

Name of the country	Case fatality ratio
Italy	14.5%
UK	14.0%
Mexico	12.0%
Spain	11.5%
Ecuador	8.4%
Indonesia	5.4%
USA	5.4%
Brazil	4.7%
Iran	4.7%
Egypt	3.9%

Source: Johns Hopkins University.

TABLE 3 Top Ten Countries with Worst Deaths per 100,000 Population

Name of the country	Deaths per 100,000 population
UK	63.99
Spain	60.60
Italy	57.19
USA	36.41
Ecuador	24.33
Brazil	23.37
Peru	23.32
Chile	21.85
Mexico	16.16
Russia	5.42

Source: Johns Hopkins University.

The economic impact of COVID-19

As the pandemic has continued to impact economic activities worldwide; as of June, the global growth projections have been further revised downward. As per the June 2020 Global Economic Prospects report released by the World Bank,[38] the global economy may shrink by as much as 5.2% in 2020, which is the deepest global recession observed since World War II. The report also cautioned that the recession could be even steeper if it takes longer to control the pandemic. In such a scenario, global growth could shrink even by 8% in 2020.

The World Bank's projections suggest that most economies will experience a recession in 2020. The advanced economies, which have observed sharp falls in economic activity, may shrink by 7%, while emerging market and developing economies may contract by 2.5%. The US GDP is expected to contract by 6.1%, and the output in the euro area is expected to contract by 9.1% in 2020. The Indian economy will shrink by 3.2%, while China may see a positive growth of 1%.

South Asia will contract by 2.7%, sub-Saharan Africa by 2.8%, the Middle East and North Africa by 4.2%, Europe and Central Asia by 4.7% and Latin America by 7.2%. Such contraction will deeply hamper the progress made by countries over the years and will take many people towards extreme poverty.[39]

The OECD has also drawn a similar picture for the global economy in its latest Economic Outlook, published in June. The organization has provided two forecasts for global growth. In the first scenario, the OECD has assumed that if there is a second wave of COVID-19 infections, the world economy will shrink by 7.6%, while in the second scenario, if the second wave of infection is avoided, the global economy will shrink by 6% but will recover by the end of 2021.[40]

Management of COVID-19

Investigation and management of COVID-19 patients has undergone several changes in late spring 2020.[41]

Investigation

Chest CT scans have emerged as a crucial investigation technique that reveals ground glass opacities that may be found even before the appearance of symptoms. Similarly, elevations in IL-6 (> 40–100), C-reactive protein (> 10× normal) and ferritin (> 1000) suggest a picture like cytokine release syndrome and impending acute respiratory distress syndrome (ARDS).

To diagnose COVID-19, rapid molecular tests are now offered (GeneXpert Cepheid < 45 min, ID NOW COVID-19 Abbot < 15 min). Reverse transcription loop-mediated isothermal amplification (RT-LAMP) assays are an emerging test to detect SARS-CoV-2 viral RNA.[42]

Treatment

Clinicians should be aware that loss of smell and loss of taste are important and common symptoms of COVID-19, and the presence of these symptoms strongly indicates a diagnosis of COVID-19. Similarly, clinicians must be aware of silent hypoxia, where the oxygen saturation can drop down and precipitate respiratory failure without any obvious signs of respiratory distress.[43] Measuring oxygen saturation with a pulse oximeter has become the first step in managing COVID-19.

The British NHS has recommended the use of ibuprofen, which was earlier not approved by certain authorities.[44] The NHS has changed its advice on only taking paracetamol to treat the symptoms of COVID-19 and now says that patients can take either paracetamol or ibuprofen to self-medicate for symptoms, such as fever or headache.[45] Recommended target SPO_2 is ≥ 94% for those who require emergency airway management and oxygen therapy. Venous thromboembolism prophylaxis should be started with low molecular weight heparin for acutely ill hospitalized adult patients.

Remdesivir

The US FDA has made remdesivir available by emergency use authorization (EUA) for treatment.[46] The drug is given at a dose of 200 mg IV on day 1 and then 100 mg IV for four more days. This dose is recommended for those in whom oxygen saturation is ≤ 94%. The medicine should be given before mechanical ventilation. The median time to recovery is reduced by 31%.

Immunomodulator

Tocilizumab, an anti-IL6R agent, is being used to counter cytokine storm.[47] The dose is 8 mg/kg IV as a single dose. It should be used early in the course of illness when IL-6 level becomes high and before intubation and the development of ARDS. Anecdotal reports from large centres with experience suggest some patients undergo rapid improvement, with improved oxygenation often within 24–48 hours of administration.

Convalescent plasma

Convalescent plasma therapy has been proposed as a useful treatment.[48] The FDA has authorized its use as an Emergency Investigational New Drug in severe or life-threatening cases of COVID-19. It has benefited patients who have ARDS. Monoclonal antibody therapy may become an alternative to plasma therapy in the near future.

Low-dose dexamethasone

The RECOVERY trial has shown that a low dose (6 mg once per day, either by mouth or by intravenous injection) of dexamethasone for ten days has reduced mortality in patients who require oxygen therapy.[49] Dexamethasone reduced deaths by one-third in ventilated patients and by one-fifth in other patients receiving oxygen only. There was no benefit for those who did not require oxygen therapy.

COVID-19 vaccine

AZD1222 (formerly ChAdOx1 nCoV-19) is among the forerunners in the development of an effective vaccine against COVIID-19. Developed by Oxford University, it is undergoing a phase 2b/3 trial, enrolling more than 10,000 human volunteers.[50] Pharmaceutical giant AstraZeneca has entered into collaboration with the Coalition for Epidemic Preparedness Innovations (CEPI); GAVI, the Vaccine Alliance; and the Serum Institute of India to produce and distribute the vaccine by the end of this year.[51] AstraZeneca reported that it has doubled manufacturing capacity for its potential coronavirus vaccine to 2 billion doses and plans to provide 400 million doses before the end of 2020. The company further stated that it has already agreed to supply 300 million doses of the potential vaccine to the United States and a further 100 million doses to the UK, with the first deliveries expected in September 2020.[52]

If everything goes well, a vaccine against COVID-19 is expected before the end of 2020. However, because vaccine development is an extremely complex process, it is difficult to project how well this rollout will occur. Safety, efficacy, affordability and equitable distribution are important factors that will play an important role in halting the global spread of the pandemic.

Notes

1 Coronavirus disease 2019 (COVID-19), *BMJ Best Practice*, https://bestpractice.bmj.com/topics/en-us/3000168/aetiology, accessed on June 13, 2020
2 Coronavirus disease (COVID-19) Situation Report – 151, available at https://www.who.int/docs/default-source/coronaviruse/situation-reports/20200619-covid-19-sitrep-151.pdf?sfvrsn=8b23b56e_2, accessed on June 19, 2020
3 Global report: WHO says the Americas are centre of pandemic as cases surge, *The Guardian*, available at https://www.theguardian.com/world/2020/may/27/global-report-who-says-the-americas-are-centre-of-pandemic-as-cases-surge, accessed on June 13, 2020

4 Ibid.

5 Brazil's biggest cities start reopening as COVID-19 surges, *Reuters*, available at https://in.reuters.com/article/health-coronavirus-brazil/brazils-biggest-cities-start-reopening-as-covid-19-surges-idINKBN23I096, accessed on June 13, 2020

6 Coronavirus: How pandemic turned political in Brazil, *BBC*, available at https://www.bbc.com/news/world-latin-america-53021248, accessed on June 14, 2020

7 Wuhan residents brave queues as coronavirus mass testing begins, *The Guardian*, available at https://www.theguardian.com/world/2020/may/15/wuhan-residents-queues-corona virus-mass-testing-begins, accessed on June 13, 2020

8 China puts city of Shulan under Wuhan-style lockdown after fresh Covid-19 cases, *The Guardian*, available at https://www.theguardian.com/world/2020/may/19/china-puts-city-of-shulan-under-wuhan-style-lockdown-after-fresh-covid-19-cases, accessed on June 13, 2020

9 Beijing reimposes lockdown measures after new Covid-19 outbreak, *The Guardian*, available at https://www.theguardian.com/world/2020/jun/13/beijing-china-new-covid-19-cases-linked-to-food-market, accessed on June 13, 2020

10 Anxiety in Beijing as officials battle new coronavirus outbreak, *The Guardian*, available at https://www.theguardian.com/world/2020/jun/19/anxiety-in-beijing-as-officials-battle-new-coronavirus-outbreak, accessed on June 20, 2020

11 Global report: six US states report most ever new coronavirus cases, *The Guardian*, available at https://www.theguardian.com/world/2020/jun/17/global-report-six-us-states-report-most-ever-new-coronavirus-cases, accessed on June 20, 2020

12 Coronavirus live updates: As social distancing wanes, Cuomo warns of another lockdown, *The New York Times*, available at https://www.nytimes.com/2020/06/14/world/coronavirus-updates.html, accessed on June 15, 2020

13 Once the coronavirus epicenter in the U.S., New York City begins to reopen, *Washington Post*, available at https://www.washingtonpost.com/nation/2020/06/08/coronavirus-live-updates-us/, accessed on June 15, 2020

14 U.S. surpasses 2 million coronavirus cases, *Washington Post*, available at https://www.washingtonpost.com/nation/2020/06/11/coronavirus-update-us/, accessed on June 14, 2020

15 https://www.nytimes.com/2020/06/12/world/coronavirus-us-usa-updates.html, *New York Times*, accessed on June 14, 2020

16 Coronavirus cases fall in France despite easing of lockdown, *The Guardian*, available at https://www.theguardian.com/world/2020/jun/11/coronavirus-cases-fall-in-france-despite-easing-of-lockdown, accessed on June 13, 2020

17 Italian prosecutors question PM Conte for three hours over coronavirus response, *Reuters*, available at https://in.reuters.com/article/health-coronavirus-italy-pm/italian-prosecutors-question-pm-conte-for-three-hours-over-coronavirus-response-idINK BN23J2J8, accessed on June 13, 2020

18 Coronavirus: UK daily deaths drop to pre-lockdown level, *BBC*, available at https://www.bbc.com/news/uk-52968160, accessed on June 13, 2020

19 Germany's Covid-19 spikes present fresh challenges as lockdown lifts, *The Guardian*, available at https://www.theguardian.com/world/2020/jun/02/germany-covid-19-spikes-present-fresh-challenges-as-lockdown-lifts, accessed on June 13, 2020

20 Coronavirus pandemic: Tracking the global outbreak, *BBC*, available at https://www.bbc.com/news/world-51235105, accessed on June 20, 2020

21 Coronavirus disease (COVID-19) Situation Report – 151

22 Russia's total coronavirus cases exceed 520,000, *Reuters*, available at https://in.reuters.com/article/us-health-coronavirus-russia-cases/russias-total-coronavirus-cases-exceed-520000-idINKBN23K07Y, accessed on June 13, 2020

23 Ibid.

24 Moscow more than doubles city's Covid-19 death toll, *BBC*, available at https://www.bbc.com/news/world-europe-52843976, accessed on June 14, 2020

25 Coronavirus disease (COVID-19) Situation Report – 151

26 Ibid.

27 India's Covid-19 peak may arrive mid-Nov; paucity of ICU beds, ventilators likely: Study, *Live Mint*, available at https://www.livemint.com/news/india/india-s-covid-19-peak-may-arrive-mid-nov-paucity-of-icu-beds-ventilators-likely-study-11592136032497.html, accessed on June 14, 2020

28 Delhi surpasses Maharashtra with 30% Covid positivity rate, *Times of India*, available at https://timesofindia.indiatimes.com/videos/city/delhi/delhi-surpasses-maharashtra-with-30-covid-positivity-rate/videoshow/76477379.cms, accessed on June 20, 2020

29 Railways deploys 54 isolation coaches in Delhi to help in Covid fight – here's how they work, *The Print*, available at https://theprint.in/health/railways-deploys-54-isolation-coaches-in-delhi-to-help-in-covid-fight-heres-how-they-work/442077/, accessed on June 20, 2020

30 Delhi health minister Satyendar Jain receives plasma therapy, kept under ICU monitoring for 24 hours, *The Times of India*, available at https://timesofindia.indiatimes.com/city/delhi/delhi-health-minister-satyendar-jain-receives-plasma-therapy-kept-under-icu-monitoring-for-24-hours/articleshow/76477417.cms, accessed on June 20, 2020

31 Coronavirus: Has a second wave of infections hit Iran? *BBC*, available at https://www.bbc.com/news/52959756, accessed on June 14, 2020

32 Coronavirus pandemic: Tracking the global outbreak, BBC

33 Global report: WHO warns of accelerating Covid-19 infections in Africa, *The Guardian*, available at https://www.theguardian.com/world/2020/jun/12/global-report-who-warns-of-accelerating-infections-in-africa-but-says-severe-cases-not-going-undetected, accessed on June 14, 2020

34 New Zealand lifts all Covid restrictions, declaring the nation virus-free, *BBC*, available at https://www.bbc.com/news/world-asia-52961539, accessed on June 15, 2020

35 From celebration to dismay: the week Covid-19 re-emerged in New Zealand, *The Guardian*, available at https://www.theguardian.com/world/2020/jun/19/from-celebration-to-dismay-the-week-covid-19-re-emerged-in-new-zealand, accessed on June 20, 2020

36 Mortality analyses, Johns Hopkins University, available at https://coronavirus.jhu.edu/data/mortality, accessed on June 20, 2020

37 Ibid.

38 Global outlook: Pandemic, recession: The global economy in crisis, *World Bank*, available at https://openknowledge.worldbank.org/bitstream/handle/10986/33748/211553-Ch01.pdf, accessed on June 20, 2020

39 The global economic outlook during the COVID-19 pandemic: A changed world, *World Bank*, available at https://www.worldbank.org/en/news/feature/2020/06/08/the-global-economic-outlook-during-the-covid-19-pandemic-a-changed-world, accessed on June 20, 2020

40 Global economy faces a tightrope walk to recovery, *OECD*, available at https://www.oecd.org/economy/global-economy-faces-a-tightrope-walk-to-recovery.htm, accessed on July 4, 2020

41 Coronavirus COVID-19 (SARS-CoV-2), by Paul G. Auwaerter, Johns Hopkins Medicine, available at https://www.hopkinsguides.com/hopkins/view/Johns_Hopkins_ABX_Guide/540747/all/Coronavirus_COVID_19__SARS_CoV_2, accessed on June 13, 2020

42 Coronavirus disease 2019 (COVID-19), *BMJ Best Practice*, available at https://bestpractice.bmj.com/topics/en-us/3000168/treatment-algorithm, accessed on June 13, 2020

43 Ibid.

44 Ibid.

45 Julia Robinson, Paracetamol or ibuprofen can be used to treat the symptoms of COVID-19, says NHS England, available at https://www.pharmaceutical-journal.com/news-and-analysis/news/paracetamol-or-ibuprofen-can-be-used-to-treat-the-symptoms-of-covid-19-says-nhs-england/20207906.article?firstPass=false, accessed on July 2, 2020

46 Coronavirus COVID-19 (SARS-CoV-2), by Paul G. Auwaerter, Johns Hopkins Medicine.

47 Ibid.

48 Ibid.

49 Low-cost dexamethasone reduces death by up to one third in hospitalised patients with severe respiratory complications of COVID-19, RECOVERY trial, available at https://www.recoverytrial.net/news/low-cost-dexamethasone-reduces-death-by-up-to-one-third-in-hospitalised-patients-with-severe-respiratory-complications-of-covid-19, accessed on June 20, 2020

50 Draft landscape of COVID-19 candidate vaccines, *WHO*, available at https://www.who.int/publications/m/item/draft-landscape-of-covid-19-candidate-vaccines, accessed on June 21, 2020 According to CDC during Phase I, small groups of people receive the trial vaccine. In Phase II, the clinical study is expanded and vaccine is given to people who have characteristics (such as age and physical health) similar to those for whom the new vaccine is intended. In Phase III, the vaccine is given to thousands of people and tested for efficacy and safety.

51 Agreements with CEPI and Gavi and the Serum Institute of India will bring vaccine to low and middle-income countries and beyond, Astra Zeneca, available at https://www.astrazeneca.com/media-centre/press-releases/2020/astrazeneca-takes-next-steps-towards-broad-and-equitable-access-to-oxford-universitys-covid-19-vaccine.html, accessed on June 21, 2020

52 AstraZeneca doubles capacity for potential Covid-19 vaccine to 2bn doses, *The Guardian*, available at https://www.theguardian.com/business/2020/jun/04/astrazeneca-doubles-capacity-for-potential-covid-19-vaccine-to-2bn-doses, accessed on June 21, 2020

INDEX

Printed in the United States
by Baker & Taylor Publisher Services

Printed in the United States
by Baker & Taylor Publisher Services